T0142841

Springer Series on Polymer and Composite Materials

Series editor

Susheel Kalia, Army Cadet College Wing, Indian Military Academy, Dehradun, India

More information about this series at http://www.springer.com/series/13173

Rasel Das
Editor

Polymeric Materials for Clean Water

Editor
Rasel Das
Chemical Engineering
Leibniz Institute of Surface Engineering
Leipzig, Sachsen, Germany

ISSN 2364-1878 ISSN 2364-1886 (electronic)
Springer Series on Polymer and Composite Materials
ISBN 978-3-030-13145-6 ISBN 978-3-030-00743-0 (eBook)
https://doi.org/10.1007/978-3-030-00743-0

Softcover re-print of the Hardcover 1st edition 2019

This Springer imprint is published by the registered company Springer Nature Switzerland AG
The registered company address is: Gewerbestrasse 11, 6330 Cham, Switzerland

Preface

The main purpose of this book entitled, *Polymeric Materials for Clean Water*, is to compile different approaches, which are used for water treatment and purification. Most of the polymer researchers have found difficulties to search for information in the ever-expanding polymer literatures published on the topic of uses of polymer for a range of water purification technologies especially coagulation and flocculation, adsorption, catalysis, disinfection, and filtration. Since scientists have experienced to locate a range of polymers used either as polymers themselves or often in the form of composites that is buried in a long journal article, the contributors to this book have compiled most of the important information based on searching the literature that polymer chemists, polymer physicists, polymer engineers, material scientists, nanotechnologist, water specialists, and environmental technologists are likely to need. Chapter 1 introduces basic chemistry of polymers and summarizes the following chapters of this book. For example, Chap. 2 describes the general methods of polymer synthesis, and the advantages, possibilities, and drawbacks of each method are discussed. Most of the modern methods of controlled polymer synthesis leading to well-defined polymers with desired structure, composition, and properties are nicely illustrated. Chapter 3 contains prominent polymer characterization techniques and the physicochemical properties of polymers. Albeit this book emphasizes on water purification methods mediated by polymers, we have added these in Chaps. 2 and 3 as they are important to understand by polymer research community. Chapter 4 brings a range of polymers necessary to improve the operational efficiency of coagulation and flocculation. Chapter 5 has focused on polymer and polymer-based nanocomposite for adsorption of water pollutants. Most of the polymer-based water purification catalysts are discussed in Chap. 6. An extensive look has been taken to discuss most of the antimicrobial polymers used for water disinfection in Chap. 7. Chapter 8 contains useful information of the uses of polymeric materials for membrane development, necessarily applied in water filtration. I am grateful if our contributors and readers send me any new information they accumulate in the course of their research, and any errors, misprints, omissions, and other flaws which are required for future editions of this book. I would like to thank all of the contributors to this book for their help and continued patience. The staffs at Springer have provided excellent help and support in getting all the work done, and I am grateful to them. I hope that the outstanding efforts of all these people will find due appreciation among the users of this book.

Leipzig, Germany Rasel Das

Contents

Chapter 1
Introduction

Rasel Das

Abstract Polymer can be defined as a macromolecule, and due to its broad range of properties, they could play an essential role in everyday life. This chapter defines some terms of polymer and its chemistry for chain formation, mechanism of polymerization, methods for polymer characterization, etc. It also highlights the summary of each chapter of this book, i.e., the uses of both natural and synthetic polymers for water purification.

Keywords Polymer · Definition · Syntheisis · Characterization
Water purification

1.1 Background

Most of the readers, especially the students, often confuse about the uses of term polymer, polymer chain, and macromolecule. Polymer is a class of large organic molecule. Alternatively, one can isolate them as a mixture of macromolecules with molecular weight is >1000 g/mol. Macromolecule consists of repeated units whose chemical structure is directly or closely linked to the monomers [1]. Many positive/negative or both charge-containing monomers are assembled in a chain to form their structure. The length of the polymer's chain varies significantly. Most of the polymers consist of thousands to millions of monomer units, and the length is between 400 and 8000 ft. Based on charge of the monomers, a polymer can be cationic, anionic, amphoteric (cationic and anionic), and nonionic. The amount of charged monomers determines the polymer overall charge density and is usually expressed by percentage. Based on chain orientation, polymers can be classified as linear, branched, and cross-linked as shown in Fig. 1.1. The chain morphology determines based on polymerization reactions. Polymerization starts by the addition of monomer molecules on an active propagating center through the formation of

R. Das (✉)
Leibniz Institute of Surface Engineering, Leipzig, Germany
e-mail: raselgeneticist@gmail.com

R. Das (ed.), *Polymeric Materials for Clean Water*,
Springer Series on Polymer and Composite Materials,
https://doi.org/10.1007/978-3-030-00743-0_1

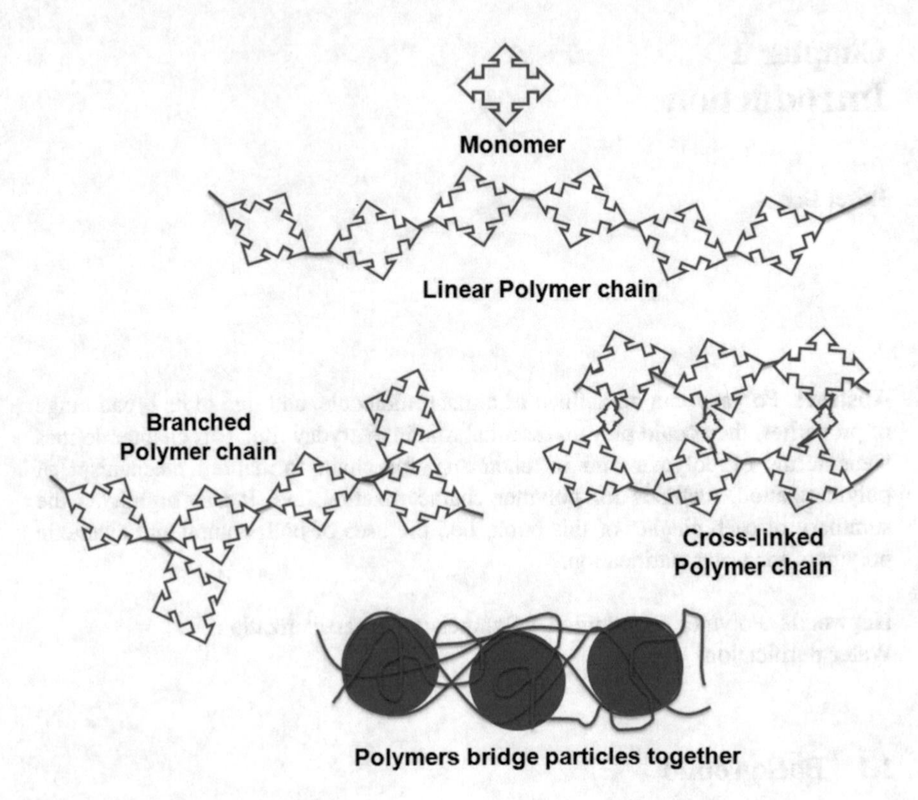

Fig. 1.1 Classification of polymers

successive new chemical bonds to produce a high molecular weight chain (i.e., macromolecule). Polymerization is a multistep procedure, so that the chain length in each polymer is not uniform. The degree of polymerization typically corresponds to the number of monomer units in a macromolecule.

Although "*n*" in Fig. 1.2 corresponds to a number of structural units per polymer chain, it is not always matched to the number (*n*) that is actually employed in a macromolecule. It depends on the chemical structure of the monomer unit.

The mechanism of polymerization is extensively discussed in Chap. 2. It includes a range of polymerization reactions, e.g., radical polymerization, controlled/living radical polymerization, ionic polymerization—anionic polymerization, living anionic polymerization, cationic polymerization, and living cationic polymerization—stereo regulation in radical and ionic polymerization, coordination polymerization, ring-opening polymerization, and multimode polymerization. Among these, the main methods of polymerization typically used in industry are radical, ionic (cationic and anionic), and coordination polymerizations. The other techniques of polymerization such as ring-opening polymerization, living radical polymerization are discovered in recent years and are promising methods for wide industrial applications and obtaining of novel materials.

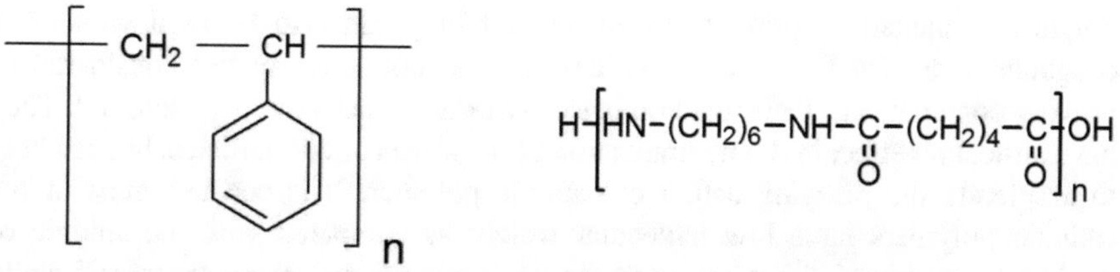

PA-6,6: $X = 2n$

Polystyrene: $X = n$

Fig. 1.2 Schematic representation of a polymer showing the two examples of the differences between the degree of polymerization and the number *n*. Figure is adapted with permission from [1]

Chapter 3 describes a range of characterization methods including mass spectrometry (MS), matrix-assisted laser desorption/ionization (MALDI)-time-of-flight (TOF)-MS, gel permeation chromatography (GPC), X-ray diffraction (XRD), Fourier transform infrared spectroscopy (FTIR), nuclear magnetic resonance (NMR), and X-ray photoelectron spectroscopy (XPS). Among them MS, GPC, and NMR are prominent characterization techniques used for analyzing polymeric materials. These are used for measuring mass-to-charge ratio (m/z) of analyte ions. XRD is helpful for solid-state analysis, e.g., degree of crystallinity and crystal structure as well as the unit cell parameters. FTIR can be used for identifying the polymer functional groups, while XPS provides information regarding the chemical composition of polymeric materials. NMR provides information on the mobility of their molecules. This chapter also highlights the physical properties of polymers including their hydrophobicity and hydrophilicity states, functional groups, and flexibility of the polymer chain structure. On the other hand, chemical properties, especially chemical reactivity, toxicity, biocompatibility, chirality, adsorption capacities, chelation, and polyfunctionality of polymeric materials, are also discussed.

A wide variety of cationic and anionic polymers are used for water clarification (WC). WC is a process of water treatment where a range of suspended particles, especially stable colloids, is removed from surface water. Whatever the purposes of water purification, e.g., drinking, agricultural, industrial, WC is a must procedure for water treatment. Most of the colloids present in surface water carry electric charge potential. Due to this same charge potential, the colloids repulse each other and floating throughout the water. It is difficult to form a large particle, which typically formed upon agglomeration, and is difficult to precipitate from the bulk solution phase. Therefore, a range of polymers is used as a coagulant, which neutralizes these colloid charges, forming aggregated colloids followed by precipitation. Coagulation is the first step of WC, but the aggregated colloids (often called mico-flocs) are too small to precipitate. So, a continuous stirring speed is maintained to facilitate further collisions of mico-flocs to form a larger floc aggregate to settle down in a basin. This process is called flocculation. Diallyldimethylammonium chloride (*DADMAC*) is a cationic flocculant organic polymer which has appreciable charge density for coagulation. The cationic polymer is named as primary coagulant, but they can also be used with other coagulant aids. On the other hand, anionic and nonionic are not considered as primary coagulants as their function basically as flocculants or coagulants aid. They have almost no effect in the neutralization of negative colloid particles, but can help to accelerate the bridging action of cationic polymer. It is because most of the cationic polymers have low molecular weight as compared with the anionic or nonionic polymers. Therefore, anionic or nonionic polymers are significantly important along with cationic polymers, especially to treat industrial WC. Due to uncoiled chain structure, which is exposed to a greater number of separate particles when added to the water, anionic polymer can accelerate the best bridging action. They can work as a primary coagulant when the colloidal particles are positively charged. Using anionic or nonionic polymers for accelerating the action of cationic polymers is cost-effective and overall could reduce treatment costs considerably. Chapter 4 discusses in details about the uses of polymers for coagulation and flocculation in wastewater treatment plant. Both the synthetic and natural polymers are extensively discussed over there. Synthetic chemical coagulants are based on organic polymers (or polyelectrolytes), while the natural-based organic coagulants are derived from both plants and animals. Their mechanism of action in the processes of coagulation and flocculation is discussed extensively.

Polymeric adsorbents and their derivatives (e.g., polymer-based composite) have been developed for over seven decades and extensively used in polluted water treatment, such as organic wastewater treatment and advanced removal of heavy metals. Chapter 5 focuses on polymers that can be used as adsorptive materials in the field of water treatment. This chapter describes in detail the uses of these promising materials for the adsorption of various organic and inorganic pollutants from contaminated waters, in terms of strong mechanical strength, excellent hydraulics performance, high stability, and tunable surface chemistry. Therefore, the physicochemical structure of polymeric materials determines the rate of adsorption of target pollutant. In short, the skeleton chemistry, pore structure, surface functional groups, and the encapsulated moieties control the maximum adsorption capacity of an adsorbent. This chapter also highlights how to synthesize a polymeric adsorbent and its nanocomposite, mechanism of action, used for the removal of various organic and inorganic pollutants.

Chapter 6 focuses on the uses of polymer-based photocatalysts for water purification. Some photosensitizers and hybrid photocatalysts along with their properties and potential applications in degradation of water pollutants have been presented either in pure or in composite forms. Since polymeric and hybrid photocatalysts are especially well suited for the removal of chemical compounds, this chapter extensively discusses them.

Since microbial infection appeared as one of the most critical environmental pollutions, water security has urged both researchers and industries to develop cost-effective antimicrobial polymer system. Water disinfection using polymer is a sustainable water treatment process. Among all the antimicrobial polymers, the use of chemically modified chitosan or metal-complexed chitosan composites has been used extensively in the field of water disinfection. This is due to its nontoxicity effect in nature, biocompatibility, and biodegradability. Moreover, using chitosan biopolymer could enhance the efficacy of some existing antimicrobial agents, antifungal agents, and minimize the environmental problems. Therefore, Chap. 7 describes all the antimicrobial polymers in the beginning including poly[1,3-dichloro-5-methyl-5-(4′-vinylphenyl) hydantoin] (poly1-Cl) and poly[1,3-dibromo-5-methyl-5-(4′-vinylphenyl) hydantoin] (poly1-Br), polyamide, poly-(*N*-benzyl-4-vinylpyridinium bromide) (PVP), polymeric silsesquioxanes chloride, polyepicyanuriohydrin, poly (glycidyl methacrylate), but later focuses on the chitosan or its composites, particularly synthesis, properties, and demonstration of water disinfection. Antimicrobial polymers are typically used as a tertiary method of treatment in processes like polymer-assisted filtration to solve our current water disinfection issue.

Due to more flexibility and less expensive material, many polymers have been used for membrane fabrication and its possible uses in the filtration of organic, inorganic, biological pollutants, etc. These membrane systems classify into reverse osmosis (RO), nanofiltration (NF), ultrafiltration (UF), microfiltration (MF) processes. The polymers used in these four processes include cellulose acetate (CA), polyamide (PA), polyvinylidene fluoride (PVDF), polysulfone (PSF), polyethersulfone (PES), polyvinyl chloride (PVC), polyimide (PI), polyacrylonitrile (PAN), polyethylene glycol (PEG), polyvinyl alcohol (PVA), poly(methacrylic acid) (PMAA), poly(arylene ether ketone) (PAEK), poly(ether imide) (PEI), polyaniline nanoparticles (PAN), and polyethersulfone amide (PESA). Chapter 8 presents these all types of polymers, their uses for membrane casting, and demonstration for their uses in water purification. Some polymers-based filters suffer some drawbacks, such as chemical attack of polymers, membrane fouling, and hydrophobicity. So, this chapter highlights some nanocomposites, i.e., polymers mixed with nanomaterials through blending and surface modification techniques. All of these methods are discussed in detail and also highlight further improvements of polymeric membrane for developing third-generation filtration techniques.

References and Future Reading

1. Hamaide T, Holl Y, Fontaine L, Six J-L, Soldera A (2012) Teaching polymer chemistry: revisiting the syllabus. Open J Polym Chem 2:132

References and Further Reading

Chapter 2
Mechanisms of Polymer Polymerization

Dmitry F. Grishin and Ivan D. Grishin

Abstract This chapter is devoted to the observation of general methods of polymer synthesis including radical, ionic, coordination, and metathesis polymerization. The main advantages, possibilities, and drawbacks of each method are discussed. A special emphasis was placed to the modern methods of controlled polymer synthesis leading to well-defined polymers with desired structure, composition and properties. Such methods are considered as a way to the novel polymer materials for various high-tech applications.

Keywords Polymerization methods · Modern techniques · Controlled synthesis

2.1 Introduction

The development of novel high-tech industries is impossible without development of new materials with desired and predetermined properties and parameters. Synthetic polymers occupy a special place among such materials. Nowadays, polymers often replace metals and other natural materials in many applications. For example, they are widely used as constructing materials due to its high plasticity in conjunction with high strength, low density, durability, and stability toward environment. For example, more than 50% of materials used for production of a newest Boeing 787 Aircraft are plastics and composites. The same thing is observed in automotive industry. Fenders and other parts of car bodies produced by leading brands are made from plastics. The leading experts in machinery construction agree in opinion that the use of plastics as constructive materials will increase in near future [1].

Polymerization is a process of forming compounds with high molecular weight through the consistent addition of monomer molecules to active propagating center.

D. F. Grishin (✉) · I. D. Grishin
Lobachevsky State University of Nizhny Novgorod, Nizhny Novgorod, Russia
e-mail: grishin@ichem.unn.ru

R. Das (ed.), *Polymeric Materials for Clean Water*,
Springer Series on Polymer and Composite Materials,
https://doi.org/10.1007/978-3-030-00743-0_2

The monomers are generally small molecules which contain unsaturated bond or cyclic fragments. A general scheme of formation of macromolecules from monomer units may be described by the following equations:

$$n\ \mathrm{A{=}B} \longrightarrow \left[\mathrm{A{-}B}\right]_n \quad (2.1)$$

$$n\ \underset{\mathrm{C}}{\mathrm{A{-}B}} \longrightarrow \left[\mathrm{A{-}B{-}C}\right]_n \quad (2.2)$$

In this scheme *n*, a number of structural units per polymer chain are called a degree of polymerization and serve as a merit of a polymer chain length.

Among the main methods of polymerization used in industry, the most essential ones are radical, ionic (cationic and anionic), and coordination polymerizations. The novel techniques of polymerization such as ring-opening polymerization, living radical polymerization discovered in recent years are promising methods for wide industrial applications and obtaining of novel materials. The main regularities of different ways of polymerization in different conditions are described below.

2.2 Radical Polymerization

Radical polymerization is among the most important and widely used methods for industrial production of polymers. Nowadays, about 50% of all polymers made by chemical industry are produced by this method. A wide range of monomers is capable to react via radical mechanism forming polymers with valuable properties. Among them are ethylene, vinyl chloride, vinyl acetate, vinylidene chloride, tetrafluoroethylene, acrylonitrile, methacrylonitrile, methyl acrylate, methyl methacrylate, styrene, butadiene, chloroprene, and many others.

The mechanism of radical polymerization is a chain radical process consisted of initiation, propagation, termination, and transfer reactions as key stages [2, 3].

(a) **Initiation**

It lies in generation of initial propagating chains and may proceed under thermal, photochemical, radiation, or chemical impact. A reactive species of radical nature capable for reaction with monomer appears in reaction media during this process. The most common and simple way of initiation is based on conducting of thermal decomposition of special compounds in the polymerization media. Such compounds capable to generate active radicals under thermal treatment are called initiators. The general scheme of initiation may be represented by two stages. The first one is decomposition of initiator giving primary radicals ($R^{\cdot}$) which interact with monomer (M) with formation of propagating chains:

$$I \rightarrow 2R^{\cdot} \quad (2.3)$$

$$R^{\cdot} \xrightarrow{+M} RM^{\cdot} \quad (2.4)$$

The concentration of initiating radicals $R^{\cdot}$ depends on [I], the rate constant of dissociation k_d, and initiation efficiency f. The initiator's efficiency itself depends on the ability of $R^{\cdot}$ to add to the monomer instead of undergoing termination. Thus, the efficiency depends on monomer concentration and the ratio k_i/k_p, where k_i is the rate constant of the initiation reaction and k_p is the rate constant of the propagation reaction (see later). Only, some radicals $R^{\cdot}$ succeed in the initiation step to form $RM^{\cdot}$, which becomes the precursor of a polymer chain. Initiating radicals are generated from a compound I with kinetics governed by k_d.

Among widely used initiators are different peroxides including alkyl peroxides (e.g., tert-butyl peroxide), hydroxyl peroxides (e.g., tert-butyl hydroperoxide, cumene hydroperoxide), peroxoesters (tert-butyl peroxobenzoate), peroxo anhydrides (benzoyl peroxide). Peroxides are capable to decompose under heating or irradiation in accordance with scheme:

$$Ph{-}\underset{\underset{O}{\|}}{C}{-}O{-}O{-}\underset{\underset{O}{\|}}{C}{-}Ph \longrightarrow 2\,Ph{-}\underset{\underset{O}{\|}}{C}{-}O^{\cdot} \longrightarrow 2\,Ph^{\cdot} + 2\,CO_2 \quad (2.5)$$

Diazocompounds are also widely employed as a source of radicals for polymerization. Among them, the azobisisobutyronitrile (AIBN) is the most commonly used one. Its decomposition results in eliminating of nitrogen molecule and formation of two carbon-centered radicals:

$$H_3C{-}\overset{\overset{CH_3}{|}}{\underset{\underset{CN}{|}}{C}}{-}N{=}N{-}\overset{\overset{CH_3}{|}}{\underset{\underset{CN}{|}}{C}}{-}CH_3 \longrightarrow 2\,H_3C{-}\overset{\overset{CH_3}{|}}{\underset{\underset{CN}{|}}{C}}{\cdot} + N_2 \quad (2.6)$$

The choice of proper initiator is determined by temperature of polymerization and desired rate of radical generation. For example, AIBN is usually applied at 50–70 °C, benzoyl peroxide at 80–95 °C, while tert-butyl peroxide is most effective at 120–140 °C. Activating energy of initiation is close to bond dissociation energy in initiator and usually has a value near 105–175 kJ/mol. A radical formed during the decay of initiator molecule is attached to the double bond of monomer forming propagating chain:

$$R^{\cdot} + H_2C{=}CHX \rightarrow R{-}CH_2{-}\dot{C}HX \quad (2.7)$$

In case of conducting of polymerization at high temperatures, the process may proceed without specially introduced initiators. In this case, generation of radicals occurs by decomposition of low amounts of peroxides formed as a result of slow oxidation of monomer by air or other impurities. For several monomers such as styrene and its derivatives, a thermal self-initiation may take place at high temperatures. Initiation of radical polymerization at room temperature or below may be realized using redox initiating systems [4]. An oxidation–reduction reaction may be conducted in polymerization mixture as in organic so in aqueous media. A typical example of redox initiation in aqueous media is the reaction of hydrogen peroxide with ferrous ion:

$$Fe^{2+} + H_2O_2 \rightarrow Fe^{3+} HO^{\cdot} + OH^{-} \quad (2.8)$$

Another example of redox reaction used for initiation of radical polymerization in organic media is interaction of benzoyl peroxide with methylaniline:

$$CH_3{-}NH{-}C_6H_5 + Ph{-}\underset{\underset{O}{\|}}{C}{-}O{-}O{-}\underset{\underset{O}{\|}}{C}{-}Ph \longrightarrow Ph{-}\underset{\underset{O}{\|}}{C}{-}O^{\cdot} + Ph{-}\underset{\underset{O}{\|}}{C}{-}OH + CH_3{-}\dot{N}{-}C_6H_5 \quad (2.9)$$

Photochemical or photoinitiation occurs when radicals are produced by irradiation of a reaction system with ultraviolet or visible light. In this process, a generation of free radicals may proceed either by direct dissociation of molecule after light absorption or by interaction of excited molecules with a second compound. In the latter case, a photosensitizing process takes place. The advantages of photoinitiation are the possibility to turn on and off polymerization by turning the light source on and off as well as to conduct process in special predetermined areas of the sample. This method is widely used as in laboratory experiments so in industrial scale, for example for surface and other thin-layer applications, in photoimaging industry as well as in dental medicine.

Radical polymerization can also be initiated by ionizing radiation of high energy (electrons, neutrons, α-particles, gamma, and X-rays). The interactions of these radiations with matter are complex, but general direction is ionization of a compound by ejection of an electron and forming radical cations which can propagate at either the radical and/or cationic centers depending on reaction conditions or dissociate to form separate radical and cationic species.

Activation energy of radiation-initiated as well as photoinitiated polymerization is close to zero. The most frequently used radiation is electrons. A so-called electron beam technology is used in coatings, microelectronics, and other thin reaction systems. The advantage of this method lies in the greater depth of electron penetration in comparison with UV or visible light. At the same time, higher equipment cost and risk of ionizing radiations from a health–safety viewpoint limit application of this method.

(b) **Propagation**

This is a key stage of polymerization process. It consists in the consecutive addition of monomer units to the radical centers formed as a result of initiation:

$$RM^{\cdot} \overset{+nM}{\rightarrow} RM^{\cdot}_{n+1} \quad (2.10)$$

The propagation rate is determined by concentrations of propagating radicals and propagation rate constant k_p. The latter depends on monomer structure (see Table 2.1). The addition of a monomer unit results in a radical structurally similar to the radical before the addition; therefore, there is no alteration in the stability of the growing radical.

Given that the propagation involves the addition of unsymmetrical alkenes, regioselectivity is an important issue. As a general rule, the less substituted carbon is more preferred to produce a bond. However, the more substituted carbon is not totally discriminated. Thus, for a given monomer, different modes of bonding are observed and their percentages depend on the nature of substituent in the monomer. For a substituent providing a strong stabilization to the end-free radical (e.g., phenyl ring in styrene), the regioselectivity increases; that is, the head-to-tail addition (path (a) on Eq. 2.11) is predominant in comparison with the head-to-head addition (b):

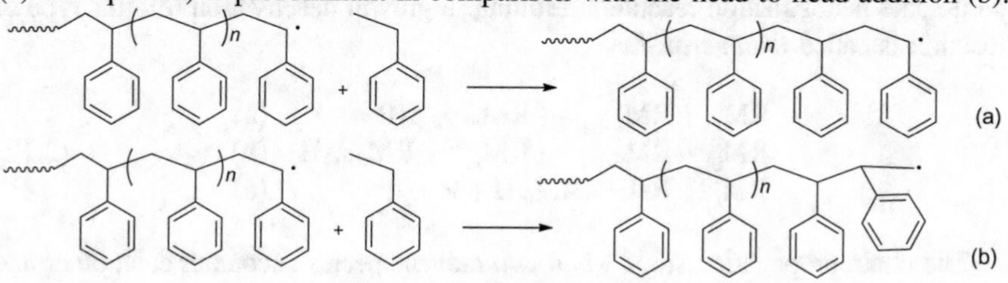

(a)

(b)

(2.11)

In the case of radical polymerization, the monomer addition is practically not stereoselective, as at the moment of the monomer additions, there is no preferred conformation of the radical. Therefore, there is no regular conformation of monomer units in the polymer chain. The sp^2 hybridization in the carbon atoms of a double bond and the resulting π-bond favor a planar arrangement of the two carbon atoms and the four immediate ligand atoms. On the other hand, the geometry of

Table 2.1 Propagation rate constants of some monomers at 25 °C [3]

Monomer	Chemical structure	K_p (l/mol s)
Vinylidene chloride	$CH_2{=}CCl_2$	9
Styrene	$CH_2{=}CH{-}C_6H_5$	35
Chloroprene	$CH_2{=}CCl{-}CH{=}CH_2$	228
Acrylic acid	$CH_2{=}CH{-}COOH$	650
Methyl methacrylate	$CH_2{=}C(CH_3){-}COOCH_3$	1010
Vinyl chloride	$CH_2{=}CHCl$	3200
Acrylamide	$CH_2{=}CH{-}C(O)NH_2$	18,000
Acrylonitrile	$CH_2{=}CH{-}CN$	28,000

alkyl radicals is considered to be a shallow pyramid (between sp^2 and sp^3 hybridization); the energy required to invert the pyramid is very small. Only the steric hindrance or electrostatic forces slightly affect the orientation of the monomer substituent, and the resulting polymer is atactic; that is, it does have a random spatial orientation of substituent near π-bond. The formation of isotactic or syndiotactic polymers is not possible during radical polymerization [3].

(c) **Termination and chain transfer reactions**

The next step of radical polymerization is termination. The rate constants of this reaction are very high ($k_t = 1 \times 10^7$ to 1×10^8 l/(mol s)), and the very low concentration of propagating chains is critical for the radical to survive some seconds or fractions of a second before the encounter with another radical species [5]. A propagating chain can be deactivated through one of the several possible reactions to become a polymer molecule. Termination is generally associated with coupling (a) and disproportionation (b) reactions, but a propagating radical can also participate in abstraction reactions resulting in growth deactivation (c); this type of reaction is called *chain transfer*.

$$\begin{array}{ll} RM_m^{\cdot} + RM_{n+1}^{\cdot} \rightarrow RM_{m+n+1}R & (a) \\ RM_m^{\cdot} + RM_{n+1}^{\cdot} \rightarrow RM_m^{-H} + RM_{n+1}H & (b) \\ RM_m^{\cdot} + RH \rightarrow RM_mH + R^{\cdot} & (c) \end{array} \qquad (2.12)$$

The chain propagation stops when two radical species encounter each other and recombine to form a larger chain or disproportionate resulting in two inactive polymer chains. A chain termination rate is diffusion controlled and is determined by how fast molecules move.

There is no effective method for the measurement of the termination rate. Major difficulties in rate constant (k_t) determination arise from the diffusion control of this reaction. Termination rate may depend on segmental and translation diffusion (and reaction–diffusion) of radical species occurring in an increasingly viscous medium that change with monomer conversion. In other words, because of the decrease of the diffusion coefficient with molecular size, the termination rate coefficient is lower at higher chain lengths. The molecular weight is self-regulated by the termination reaction and the inherent side reactions. Externally, molecular weight can be also adjusted by the reaction conditions. So, an increase of temperature causes faster radical initiator decomposition, resulting in a lower molecular weight because more chains are created. The higher pressure on the contrary increases the propagation and inhibits the termination resulting in higher MW. The increase of the radical concentration has a similar effect to that of the increase of temperature.

A propagating polymer chain can also terminate via chain transfer reaction to a special compound—transfer agent (TA). This will lead to termination of the propagating chain, along with the generation of a radical on the small molecule that can initiate another propagating chain:

$$\begin{aligned} RM_m^{\cdot} + TA &\overset{k_{trTA}}{\rightarrow} RM_{m+n+1}T + A^{\cdot} \\ A^{\cdot} + M &\overset{k_{iA}}{\rightarrow} AM^{\cdot} \end{aligned} \tag{2.13}$$

On this equation $A^{\cdot}$ is the radical resulting from the activation of the small molecule. Usually, the net effect of the chain transfer is negligible on the polymerization rate, since there is no net creation or destruction of radicals (their *nature* changes but *not* their *number*), but it causes a decrease in MW. However, other effects may occur depending on the relative values of k_{trTA} and k_{iA} with respect to k_p. The units of the rate coefficients are liter per mole per second or cubic meter per mole per second. It is a common practice to report the ratio of the transfer rate coefficient to the propagation rate coefficient, and this quantity is denoted as transfer constant $C_{trTA} = k_{trTA}/k_p$; it is defined to measure the ability of each substance to produce a chain transfer reaction.

The small species TA can be a monomer (M), a solvent (S), a chain transfer agent (CTA), an initiator (I), or an impurity (X) in the system.

Chain transfer to monomer: Table 2.2 lists values of transfer constants on monomer (C_M) for some common monomers. In the absence of other transfer reactions, chain transfer to monomer will impose an upper limit to the maximum molecular weight achievable in the polymerization of the corresponding monomer. This does not mean that this reaction will in general be the controlling step determining the molecular weight; indeed, bimolecular termination usually plays this role. High values of transfer constants to monomer are associated with high reactivity of the propagating radical.

Chain transfer to initiator: The values for chain transfer constants to initiator (Table 2.3) are generally larger than those of transfer to monomer; however, the effect of this reaction is attenuated by the fact that the initiator is present in very small amount with respect to the monomer, as the rate of transfer to initiator is k_{trI} [P][I].

Chain transfer to chain transfer agents and solvent: When polymerization takes place in a solvent, it is very important to be aware of possible chain transfer to solvent reactions. On the other hand, there are many instances, especially in industrial processes, in which it is convenient to include in the polymerization

Table 2.2 Values of chain transfer to monomer constant (C_M) for some polymerization systems [3]

Monomer	Temperature	$C_M \times 10^4$
Acrylonitrile	60	0.26–1.02
Butyl acrylate	60	0.333–1.05
Ethylene	60	0.4–4.4
Methyl methacrylate	60	0.07–0.18
Styrene	60	0.07–1.37
	70	0.6–2.0
Vinyl acetate	60	1.75–2.8
Vinyl chloride	50	8.5
	60	12.3

Table 2.3 Values of chain transfer to initiator constant for some polymerization systems [3]

Initiator	T (°C)	C_I		
		Styrene	Methyl methacrylate	Vinyl acetate
2,2′-Azobisisobutyronitrile (AIBN)	60	0–0.16	0	–
	60	–	0–0.02	0.032–0.15
Benzoyl peroxide	70	0–0.18	–	–
	80	0.13–0.813	–	–
	60	–	–	0.10
Lauroyl peroxide	70	0–0.024	–	–
	60	–	0.10–0.17	0–0.16
Palmytoil peroxide	70	0.142	–	–
	70	0.031	–	–
Tert-butyl peroxide	80	0.0027	–	–

recipe some species (so-called *chain transfer agent*—CTA) that have a high chain transfer constant. Among different compounds used for that purpose, sulfur compounds are by far the most popular chain transfer agents (Table 2.4).

The addition of a CTA is rather common, for example, in emulsion polymerization where the compartmentalized nature of the reaction tends to produce very high MW polymer due to the relative isolation of the propagating radicals in very small particles. The rates of reaction of chain transfer to a solvent (S) and to a CTA are conceptually the same: $k_{\mathrm{tr}}\mathrm{S}[\mathrm{P}][\mathrm{S}]$ and $k_{\mathrm{tr}}\mathrm{CTA}[\mathrm{P}][\mathrm{CTA}]$, respectively. The values of chain transfer constants to solvent and to CTAs are usually reported in one table, as they lie in a continuum going from low values for solvents to rather high values for CTAs (Table 2.4).

Chain transfer to polymer: During radical polymerization, intermolecular chain transfer, which involves two independent polymer chains (one active and the other dead), can take place. It is also possible that intramolecular chain transfer occurs (also called *backbiting*), in which the hydrogen abstraction occurs in the same active chain, a few carbons (about five) before the active end of the growing polymer. Intermolecular transfer will give rise to long branches, while intramolecular transfer will be the origin of short branches. Both short and long branches have a profound influence on the physical and rheological properties of the polymer formed [6].

The experimental determination of the chain transfer to polymer constant is difficult, as it does not necessarily result in a decrease of the molecular weight of the polymer. Since it involves hydrogen abstraction, the activation energy of chain transfer to polymer is relatively high (compared to propagation) [1]. Reaction conditions that favor transfer to polymer are high temperatures and high conversions (due to the high concentration of dead polymer present).

Table 2.4 Values of chain transfer to the chain transfer agents and solvent for some polymerization systems [3]

Solvent or CTA	*T* (°C)	C_s		
		Ethylene	Methyl methacrylate	Styrene
Acetone	60	–	0.195	0.32–4.1
	80	–	0.225–0.275	–
	130	160–168	–	–
iso-Butanethiol	60	–	0.66–067	21.0–25.0
	80	–	–	17.0
	130	5.8	–	–
Carbone tetrachloride	60	–	0.42–20.11	69–148
	80	–	2.4–24.4	133
	140	1.600–180.000	–	–
Chloroform	60	–	0.454–1.77	0.41–3.4
	80	–	1.129–1.9	0.50–0.916
	140	3.210–37.600	–	–
Ethyl benzene	60	–	0.766	0.67–2.7
	80	–	1.311–2.1	1.07–1.117
Hexane	100	–	–	0.9
	130	68	–	–
Toluene	60	–	0.17–0.45	0.105–2.05
	80	–	0.292–0.91	0.15–0.813
	130	130–180	–	–

(d) Inhibition of radical polymerization

An inhibitor is used to completely stop the conversion of monomer to polymer produced. For example, some other compounds, such as phenols, quinones, or hydroxyquinones, or even molecular oxygen, are also employed to inhibit the polymerization. The mechanism of action of these compounds involves the transformation of the propagating radical to an oxygen-centered radical that is unable to initiate polymerization:

$$\sim\!\!\sim M_n^{\bullet} + O{=}\text{C}_6\text{H}_3\text{R}{=}O \longrightarrow \sim\!\!\sim M_n{-}O{-}\text{C}_6\text{H}_3\text{R}{-}O^{\bullet} \quad (2.14)$$

Some stable radicals, such as diphenylpicrylhydrazyl and 2,2,6,6-tetramethyl piperidinyloxyl (TEMPO), can be used as inhibitors of the radical polymerization:

$$\sim\!\!\sim M_n^{\bullet} + (C_6H_5)_2N{-}\dot{N}{-}C_6H_2(NO_2)_3 \longrightarrow (C_6H_5)_2N{-}N(M_n)(C_6H_2(NO_2)_3) \quad (2.15)$$

(e) **Kinetics of radical polymerization**

The rate of initiation in case of using thermal decomposing initiator I may be expressed by equation:

$$V_\mathrm{i} = 2fk_\mathrm{d}[\mathrm{I}] \tag{2.16}$$

[I] is initiator concentration, f initiation efficiency which is generally equal to 0.5–1.0, and k_d constant of dissociation.

In assumption that reactivity of propagating radicals does not depend on its degree of polymerization, the propagation rate may be expressed as:

$$V_\mathrm{p} = k_\mathrm{p}[\mathrm{R}^\cdot][\mathrm{M}] \tag{2.17}$$

where k_p is constant of propagation, [R] concentration of propagating radicals, and [M] monomer concentration.

The rate of termination via bimolecular interaction leading to recombination or disproportionation may be represented as:

$$-\frac{\mathrm{d}[\mathrm{R}^\cdot]}{\mathrm{d}t} = k_0[\mathrm{R}^\cdot]^2 \tag{2.18}$$

where k_0 is termination constant (in assumption that it does not depend on chain length).

In case of the absence of side reactions, the overall rate of polymerization is equal to monomer consumption. So, we can conclude that:

$$-\frac{\mathrm{d}[\mathrm{M}]}{\mathrm{d}t} \equiv V_\mathrm{p} = k_\mathrm{p}[\mathrm{R}^\cdot][\mathrm{M}] \tag{2.19}$$

If polymerization proceeds without inhibitor or other molecules reacting with radicals, the latter decay only by recombination or disproportionation, so:

$$\frac{\mathrm{d}[\mathrm{R}^\cdot]}{\mathrm{d}t} = V_\mathrm{i} - k_0[\mathrm{R}^\cdot]^2 \tag{2.20}$$

As the radical concentration is very low ($\sim 10^{-8}$M) and is very hard to be determined, this term should be eliminated from equation. In order to do this, the steady-state assumption is made that the concentration of radicals increases initially, but almost instantaneously reaches a constant, steady-state value. In accordance with this assumption, the rate of change of the concentration of radicals remains zero during the course of the polymerization. Propagating of radical polymerization fulfills the made assumption after a few seconds after beginning of reaction. In this case:

$$V_{\mathrm{i}} = k_0[\mathrm{R}^{\cdot}]^2 \tag{2.21}$$

$$[\mathrm{R}^{\cdot}] = \sqrt{\frac{V_{\mathrm{i}}}{k_0}} \tag{2.22}$$

$$-\frac{\mathrm{d}[\mathrm{M}]}{\mathrm{d}t} = k_{\mathrm{p}}\sqrt{\frac{V_{\mathrm{i}}}{k_0}}[\mathrm{M}] \tag{2.23}$$

Degree of polymerization is equal to the average number of monomer units introduced into one polymer chain and characterizes the average molecular weight of polymer sample. It is determined as a ratio of monomer molecules consumed during polymerization to the number of formed polymer chains. At low monomer conversions when monomer concentration is almost permanent and chain transfer to polymer is negligible, it may be expressed as:

$$\overline{P_n} = \frac{V_{\mathrm{p}}}{V_{\mathrm{p}} + \sum V_{\mathrm{tr}}} \tag{2.24}$$

where V_0 is rate of bimolecular termination and $\sum V_{\mathrm{tr}}$ is sum of all transfer reaction rates.

Polymerization at high monomer conversions. The considered kinetic equations and regularities are true only for polymerization at initial stage of polymerization up to 10–15% conversion. A further increase of monomer conversion results in significant increase of viscosity of polymerization media which limits the diffusion, mostly of propagating chains. High viscosity of media affects termination reaction and results in its decrease up to two orders of magnitude. It leads to spontaneous increase of molecular weight of forming polymers. A so-called gel effect, which is also known as Trommsdorff or Norrish–Smith effect [7, 8], is observed.

This consists in an autoacceleration of the reaction as the conversion increases, and it is due to an effective decrease in the termination rate as the growing radicals encounter more difficulty in diffusing in the increasingly viscous medium.

Some reviews on previous gel effect models or on the concepts on which they are based have also been published [9, 10].

(f) **Radical copolymerization**

Conducting a simultaneous polymerization of several monomers in one vessel results in formation of copolymers. Such macromolecules contain different monomer units connected in certain sequence. Depending on this sequence random, alternating, gradient, grafted, block copolymers are distinguished [1].

In case of copolymerization of two different monomers, four different elemental reactions may take place in the assumption that reactivity is determined only by ultimate unit:

Propagating reaction (k_p)	Reaction rate
$\sim R_1^{\cdot} + M_1 \rightarrow \sim R_{11}^{\cdot}$ (k_{11})	$k_{11}[R_1^{\cdot}][M_1]$
$\sim R_1^{\cdot} + M_1 \rightarrow \sim R_{12}^{\cdot}$ (k_{12})	$k_{12}[R_1^{\cdot}][M_2]$
$\sim R_2^{\cdot} + M_1 \rightarrow \sim R_{21}^{\cdot}$ (k_{21})	$k_{21}[R_2^{\cdot}][M_1]$
$\sim R_2^{\cdot} + M_1 \rightarrow \sim R_{22}^{\cdot}$ (k_{22})	$k_{22}[R_2^{\cdot}][M_2]$

where **Mi** is **i**-type monomer, **~Rj** macroradical with **Mj** at the end, and **kij** constant for addition of **Mj** monomer to macroradical **~Ri**.

This equation describes so-called ultimate unite model of copolymerization. In a steady-state assumption, concentrations of propagating radicals $\sim R_1^{\cdot}$ and $\sim R_2^{\cdot}$ are constant and their rates of interconversion are equal:

$$k_{12}[R_1^{\cdot}][M_2] = k_{21}[R_2^{\cdot}][M_1] \tag{2.25}$$

The rates of disappearance of the two monomers, which are synonymous with their rates of entry into the copolymer, are given by:

$$-d[M_1]/dt = k_{11}[R_1^{\cdot}][M_1] + k_{21}[R_2^{\cdot}][M_1] \tag{2.26}$$

$$-d[M_2]/dt = k_{12}[R_1^{\cdot}][M_2] + k_{22}[R_2^{\cdot}][M_2] \tag{2.27}$$

From these equations, a ratio of monomer units in copolymer macromolecules may be written as:

$$\frac{m_1}{m_2} = \frac{d[M_1]}{d[M_2]} = \frac{[M_1]}{[M_2]} \cdot \frac{r_1[M_1] + [M_2]}{[M_1] + r_2[M_2]} \tag{2.28}$$

where $r_1 = k_{11}/k_{12}$ and $r_2 = k_{22}/k_{21}$

This equation is known as the copolymerization equation or the copolymer composition equation. The parameters r_1 and r_2 are termed the monomer reactivity ratios. Each r as defined above is the ratio of the rate constant for a reactive propagating species adding its own type of monomer to the rate constant for its addition of the other monomer. These values depend on chemical nature of both reacting monomers.

Values of r_1 and r_2 may be determined from experimental data. These values allow to predict composition of copolymer and distribution of monomer units at any composition of monomer mixture of two monomers. It should be mentioned that r values for radical polymerization weakly depend on the temperature and the nature of the solvent used for polymerization.

The properties of copolymers depend not only on its average composition but also on homogeneity of monomer unit distribution. The polymers with the same average composition but different types of monomer distributions are characterized by different properties. For example, copolymers formed from two distinct monomers with 1:1 ratio may be random, alternating (in case of strict alternation of units), or block copolymer (in case of formation of blocks of units of both types).

The composition homogeneity of polymers is a merit of deviation of composition of selected macromolecule from average composition of polymer sample. Different polymer chains are initiated at different times during a polymerization and propagate under different feed compositions as conversion progresses. This fact results in different compositions of polymer chains formed by conventional radical copolymerization. Some novel methods of polymer synthesis allow overcoming this drawback of radical polymerization. Controlled radical polymerization is among them.

2.3 Controlled/Living Radical Polymerization

One of the brightest and the most important latest advancements of the synthetic macromolecular chemistry was the discovery of the method of controlled radical polymerization (CRP). It significantly extended possibility of radical polymerization for preparation of new polymer materials. This procedure combining in itself the advantages, on the one hand, of free radical polymerization [11, 12] and, on the other hand, of living anionic polymerization [13] became nowadays the most effective instrument for the preparation of polymers with a targeted structure, composition, and properties [14, 15]. Some definite examples of the practical utilization of CRP methods for the synthesis of polymer materials in the industry are compiled in surveys [15–18].

The main concept underlining the CRP consists in the replacement of the irreversible termination reaction of the propagating polymer chain that proceeds very fast [with a rate constant 10^6–10^8 l/(mol s)] by a reversible interaction of the macroradical with a specially added agents X (Eq. 2.29) [18–20].

$$\sim P_n^{\cdot} + X \underset{k_d}{\overset{k_t}{\rightleftharpoons}} \sim P_n{-}X \qquad (+M,\ k_p) \qquad \sim P_n^{\cdot} + \sim P_m^{\cdot} \xrightarrow{k_0} \sim P_m{-}P_n\sim \tag{2.29}$$

$\sim P_m^{\cdot}$ and $\sim P_m^{\cdot}$ are growing macroradicals; k_d, k_r, k_p, k_t are reaction constants of reversible termination, reinitiation, propagation, and irreversible termination of the chain.

In this case after several acts of monomer addition, the propagating chain gets into a dormant state which conserves for some time. After that, it activates again, and the growth is resumed. The activity periods of the chain are ~1 ms, and they alternate with the dormant periods which are much longer, ~1 min [18]. Thus, the realization of the mechanism of the reversible stop of the polymerization in keeping with Eq. 2.29 actually permits increasing the time of formation of each polymer chain from several seconds to days providing wide opportunities for synthetic manipulations. In particular, this provides a possibility to obtain block copolymers from a wide range of monomers or to prepare gradient copolymers forming at the copolymerization of two monomers of a different reactivity. Another important merit of the modern CRP methods is the opportunity to obtain polymers with a narrow molecular weight distribution. As a result of the constantly occurring transition of the polymer chains from the active state to the dormant and vice versa, they grow simultaneously during the total polymerization time affording polymers of uniform molecular weight. Therewith, the polydispersity index of the synthesized polymers decreases with the growing conversion of the monomer.

Depending on the reaction underlying the transition of the active polymer chain in the "dormant" state and on the kinetic model that can describe the process, all the known ways of performing CRP may be tentatively divided into two large groups. Processes belonging to the first group are based on the reversible termination of the polymerization by the reaction of the growing radical with the specially introduced regulating agents (Eq. 2.29). In this group, the polymerization should be placed along the reversible termination mechanism [20–24] and the atom transfer radical polymerization [14, 15, 17, 18, 23]. Another way of regulating polymerization process is based on degenerative chain transfer reaction between active and dormant at this moment chains along Eq. 2.30.

$$\sim P_n^{\bullet} + X{-}P_m\sim \rightleftharpoons \sim P_n{\cdots}\overset{\bullet}{X}{\cdots}P_m\sim \rightleftharpoons \sim P_n{-}X + P_m^{\bullet}\sim$$

$$\sim P_n^{\bullet} \xrightarrow{+M,\ k_p} ;\quad \sim P_n^{\bullet} + \sim P_o^{\bullet} \xrightarrow{k_0} \sim P_o{-}P_n\sim ;\quad P_m^{\bullet}\sim \xrightarrow{+M,\ k_p} ;\quad P_m^{\bullet}\sim + \sim P_o^{\bullet} \xrightarrow{k_0} \sim P_o{-}P_m\sim \tag{2.30}$$

Such mechanism of reversible chain transfer may be realized using dithiocarbamates [19, 25], cobalt compounds [26, 27] as chain transfer agents. The similar mechanism is realized in radical polymerization in the presence of iodine compounds (a so-called iodine transfer radical polymerization) [28].

(a) **Reversible inhibition mechanism**

The regulating agent X in polymerization with reversible inhibition (Eq. 2.29) may be represented by stable radical or another compound capable of reversible reaction with free radicals, in particular, a complex of transition metal that can form covalent σ-bonds with alkyl radicals. In the latter case, such processes are called by the term organometallic-mediated radical polymerization (OMRP) [23, 24]. Polymerization proceeding along this mechanism implies using complexes of cobalt [29, 30], molybdenum [31, 32], titanium [24, 33], and some other metals [29, 34]. The main drawback of this kind of processes is the necessity to use the complexes of transition metals in a stoichiometric quantity with respect to initiator leading to

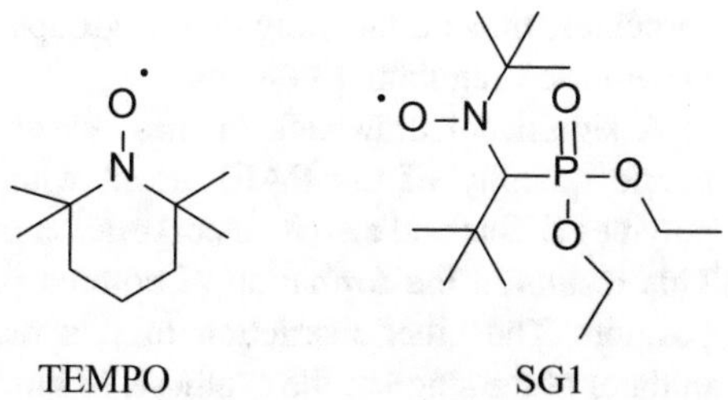

Fig. 2.1 Structures of the most commonly used agents for NMP

additional expenses due to the preparation of the agent and the purification of the polymer from it.

The majority of works in the field of the radical polymerization with reversible inhibition are describing the use as regulator (X, Eq. 2.29) converting the active chain in the dormant state the nitroxyl radicals. This type of polymerization is called nitroxide-mediated radical polymerization (NMP). First in the history, CRP agent acting by the mechanism of the reversible inhibition was the nitroxyl radical (2,2,6,6-tetramethylpiperidin-1-yl)oxyl (TEMPO) and its analogs [35]. In the last years, a large number of nitroxyl radicals of diverse structures were developed capable of acting as polymer chain growth and termination regulators. In particular, the most efficient regulators of the vinyl monomers polymerization process among the nitroxyl radicals are spin-adducts SG1 containing a phosphonate fragment in their composition [36]. Nowadays, compounds of this type are used in industry for the production of block copolymers applied as dispersants in cosmetics and other production [16] (Fig. 2.1).

(b) **Degenerative chain transfer mechanism**

Polymerization with the reversible or degenerative chain transfer known as reversible addition–fragmentation chain transfer (RAFT) [21] processes using, as a rule, organosulfur compounds as regulators [25]. In this case, the reversible equilibrium between the active and "dormant" chains is attained due to the chain transfer reaction proceeding in the polymerization system in accordance with Eq. 2.31.

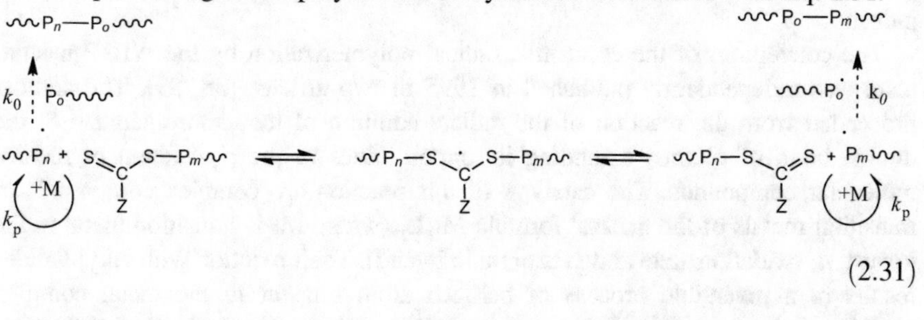

$$\sim P_n^{\bullet} + S{=}C(Z){-}S{-}P_m\sim \;\rightleftharpoons\; \sim P_n{-}S{-}\dot{C}(Z){-}S{-}P_m\sim \;\rightleftharpoons\; \sim P_n{-}S{-}C(Z){=}S + P_m^{\bullet}\sim \quad (+M,\ k_p);\quad \sim P_n^{\bullet} + P_o^{\bullet}\sim \xrightarrow{k_0} \sim P_n{-}P_o\sim;\quad \sim P_o^{\bullet} + P_m^{\bullet}\sim \xrightarrow{k_0} \sim P_o{-}P_m\sim \tag{2.31}$$

The advantage of this method consists in the high rate of the process and the very good control over molecular weight characteristics of the obtained polymers reflecting in low values of the polydispersity indices. The use of RAFT agents makes it possible to carry out the polymerization of a large number of monomers, in particular, those containing acidic groups. The latter is difficult to perform using metal-based regulating systems.

A significant drawback of these systems is the necessity of using a stoichiometric quantity of the RAFT agent with respect to the amount of the growing polymer chains and also its incorporation in the polymer chain during the synthesis. This results in the formation of colored polymers containing sulfur in their composition. The other restriction of this method is the necessity to add a radical initiator at the stage of the synthesis of block copolymers for the initial formation of polymer chains leading to formation of a certain amount of homopolymer. Besides, a certain defect is the unpleasant odor of the sulfur-containing reagents. Regardless of the mentioned shortcomings, some processes applying RAFT agents found a practical application in industry [37].

(c) **Atom transfer radical polymerization**

Another possible way of realization CRP is the use of organometallic catalyst and organohalide as initiator. Atom transfer radical polymerization (ATRP) [17, 18, 23] is based on the reversible halogen atom transfer (more seldom of other substrates) between the metal complex catalyst and the growing polymer chain (Eq. 2.32).

$$\begin{aligned} &\mathrm{R{-}Hal} \xrightleftharpoons{+\mathrm{Mt}^n\mathrm{L}_x} \dot{\mathrm{R}} + \mathrm{Mt}^{n+1}\mathrm{L}_x\mathrm{Hal} \xrightarrow{+\mathrm{M}} \dot{\mathrm{P}}_n + \mathrm{Mt}^{n+1}\mathrm{L}_x\mathrm{Hal} \\ &\mathrm{P}_n{-}\mathrm{Hal} + \mathrm{Mt}^n\mathrm{L}_x \rightleftharpoons \mathrm{Mt}^{n+1}\mathrm{L}_x\mathrm{Hal} + \dot{\mathrm{P}}_n \ (+\mathrm{M}) \xrightarrow{k_t} \end{aligned} \quad (2.32)$$

Unlike the above-discussed methods, the ATRP is a catalytic process. The latter fact makes it possible to use the metal complex in very small (catalytic) quantities with respect to the initiator and the monomer. Another important point is possibility of catalyst "tuning" for polymerization of a certain monomer. These two moments make the radical polymerization via ATRP mechanism a convenient instrument for the preparation of polymers of desired architecture both in the laboratory and in the industry.

The conception of the controlled radical polymerization by the ATRP mechanism was independently published in 1995 in two articles [38, 39]. The authors proceeded from the reaction of the radical addition of tetrachloromethane to the double bond of olefins, extending its opportunities to the production of macromolecular compounds. The catalysts of this reaction are complex compounds of transition metals of the general formula Mt_nL_x (where Mt is transition metal in the lowest, *n*, oxidation state and L is organic ligand). Their reaction with alkyl halides results in a reversible process of halogen atom transfer to the metal complex resulting in the formation in the system of alkyl radicals (Eq. 2.32). Performing this process in the monomer environment makes it possible to carry out chain propagation by fragments, i.e., to perform the radical polymerization in a controlled mode. Within the last 20 years since the first publications on this topic, many catalytic systems and compositions were developed capable of performing the polymerization by the above-described mechanism providing a possibility to synthesize polymers with a very narrow molecular weight distribution. The majority of these systems are based on the use in polymerization catalysis derivatives of copper [15, 27, 38], ruthenium [28, 39], and iron [40, 41]. Some examples of application of systems based on nickel, manganese, and other metals are also provided in the literature [23, 42].

An important achievement in the ATRP area was the development of methods permitting the significant decrease in the concentration of the applied metal catalyst. The content of residual metal amounts in the polymer in some cases is crucial for its application in microelectronics and also in biologic and medicinal technologies. Therefore, the development of CRP methods without using metal complexes is a topical trend in the field of the controlled polymer synthesis. This new direction was named metal-free atom transfer radical polymerization (MF ATRP) [43–45].

(d) **Photoinduced ATRP**

Same as in the case of the classic ATRP method, the control of the molecular weight distribution is underlain by the reversible transfer of the halogen atom between the "dormant" polymer chain and the catalyst, organic in this case. To achieve the halogen atom transfer to the molecule of the organic catalysts, its preliminary activation and transition into excited state are required. The latter is attained by irradiation with light. The subsequent reaction between the alkyl halide and the organic catalyst in the excited state can result in generation of active radicals starting polymerization process. The reverse transition of the chain in the dormant state occurs by the reaction of the propagating radical with the formed ion pair $Cat^{+}Br^{-}$. The processes occurring at MF ATRP are shown in Fig. 2.2.

Among catalysts and regulators in this type processes, various condensed aromatic derivatives were applied. Among them are anthracene, perylene, 10-phenylphenothiazine (PPT) [44], and its derivatives. Therewith among the mentioned compounds, the PPT is the most effective catalyst for the synthesis of polymers with narrow polydispersity [45].

With the use of this procedure, the polymerization in the "living chains" mode was successfully carried out for a series of acrylic monomers, including methyl

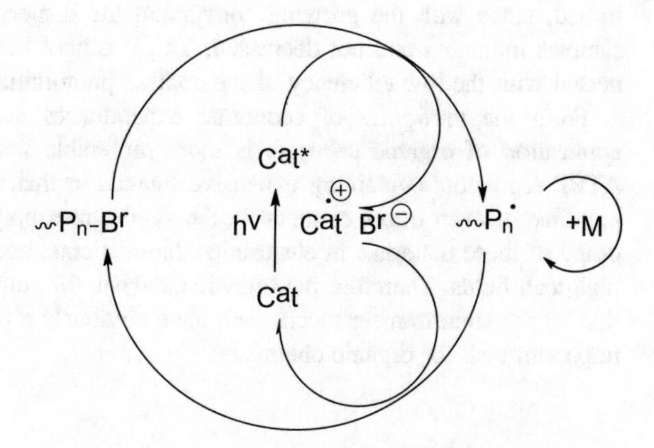

Fig. 2.2 General scheme of photoinduced ATRP mechanism

methacrylate, acrylonitrile, 2-(dimethylamino) methacrylate, and other unsaturated compounds. The polymerization process is characterized by the linear growth of the molecular weight with conversion and also with passing through the origin the linear dependence of the ratio of the log initial monomer concentration to its current concentration indicating the constant number of growing chains in the course of the polymerization. As mentioned above, it is the unambiguous proof of the polymer synthesis in the mode of growing "living chains." The range of molecular masses of polymers prepared by this method is actually analogous to this parameter for the macromolecules obtained under the conditions of classic ATRP involving the metal complexes, yet the polydispersity index in a number of instances was considerably higher showing the lower degree of the process control.

The main criterion providing the possibility to use a certain organic compound in the processes of polymerization catalysis is its ionization potential since in the course of the polymerization the catalyst first undergoes a transition in the excited state and then in a cation radical, and this process is of reversible character. For instance, the ionization potential of PPT is −2.1 eV. For comparison, the ionization potential of iridium compounds used in the processes of photoinitiated ATRP [46] has a close value of −1.7 eV. According to this parameter, not only heterocycles but also polynuclear aromatic compounds may be interesting as catalysts for MF ATRP.

Thus along with PPT and its derivatives, perylene was suggested for the use as redox-active photocatalyst for MF ATRP **4** [47]. In the presence of this catalyst polymerization afforded poly(methyl methacrylate), poly(butyl acrylate), polystyrene, and some other monomers also were polymerized with the controlled values of the molecular weight and narrow molecular weight distribution. Along with the low values of the dispersity index of the synthesized samples (M_w/M_n = 1.29–1.85), a significant confirmation of the polymerization regime of "living chains" is the presence of halogen atoms on the end of the polymer chain at the use of the bromine-containing initiators traditionally applied in the ATRP processes. At the same time, although the polymerization occurs at the constant concentration of the active centers and possesses a number of above-cited characteristics of the "living" radical polymerization, it is not possible to regard it as completely controlled, since with the growing conversion the dispersity index of the obtained samples increased and not decreased. Yet, it cannot be excluded that this is connected with the low efficiency of the applied photoinitiator [47].

From the viewpoint of economic expenditures and "green" chemistry, the application of organic catalysts is more preferable than the use of metal-based ATRP regulators containing expensive ligands in their composition. Besides, the presence of even traces of metal in the synthesized polymer essentially limits the usage of these materials in electronics, biomedicine, and in a number of the other high-tech fields. Therefore the organic catalysts for controlled radical polymerization by the atom transfer mechanism have obviously a practical interest and are an important task for organic chemistry.

The special feature of the metal-free ATRP process is the necessity to apply photoinitiation. On the one hand, like the above-mentioned case of the controlled radical polymerization by eATRP mechanism where the polymer synthesis is performed with the initiation and under electric current in an electrochemical cell, the photoinitiation can produce certain complications at the industrial implementation of this technology. On the other hand, this provides a possibility to control the course of the process and its rate by changing the light flux. The intensive development of this trend is favored by the occurring now active elaboration of new light sources based on light-emitting diodes and semiconductor lasers which are compact, highly powerful, and simple in handling.

Photocatalysis for the transmitting polymer chains from the dormant into the active state can also be used in the case of the process occurring by the mechanism of the reversible chain transfer. This CRP direction named in the literature PET for photoinduced electron transfer (PET-RAFT) is also actively developing lately [48]. It is underlain by the application of the photoinitiation to the homolysis of the C–S bond in the adducts (Eq. 2.33) leading to the formation of active radicals that in their turn carry on the polymerization process by the mechanism of reversible chain transfer.

(2.33)

The application of the photocatalysis for performing the controlled radical polymerization approaches the synthetic polymer chemistry to the processes of biopolymers syntheses in living organisms under the sunlight. The application of photocatalysis to CRP is a new important trend which later may underlie the industrial production of functional polymers, in particular, for biomedical uses [49].

(e) **Application of controlled radical polymerization in industry**

In recent years, CRP firmly acquired a place on the market of the manufactured polymer materials. Starting from the first advancements of DuPont and Daikin, within the last two decades these methods were developed in a number of other known companies including Arkema, Ciba, Kaneka, EKFA, BYK, Rhodia. As mentioned above, the methods of the controlled synthesis of macromolecules are used now in the production of new pigments, sealants, emulsion stabilizers, and block copolymers with a special complex of properties and characteristics including polymer photoresists. At the same time, the amount of commercial products manufactured under the CRP conditions remains still small as compared to the overall volume of industrially produced polymers. This is due to certain objective reasons, among which the high price of some regulators and the necessity to remove them from the polymer product should be mentioned, and also some other facts including lower polymerization rate and relatively high temperature of the process where the controlled synthesis of macromolecules is possible. As known, just the optimization of the ratio cost–quality at the manufacturing of a product is the key condition for the process implementation in industry [50]. Undoubtedly, the synergetic utilization of the experimental data on the polymerization mechanism, including the strict kinetic studies, results of computer simulation, and direct testing of catalysts would favor the targeted search for new effective metal complex systems for controlled synthesis of macromolecules under the conditions of radical initiation. In this connection, the designing of new effective catalysts, easily accessible by preparation, and convenient for application is a topical and practically important issue in the field of the controlled radical polymerization of a wide choice of monomers.

2.4 Ionic Polymerization

Radical and ionic polymerization processes have a lot in common as they proceed in a chain manner and include an initiation stage for formation of propagating centers and termination reaction giving final product. At the same time, there are some significant differences between these processes. Kinetics of ionic polymerization is usually more complex than in case of radical polymerization as reaction centers may exist in different forms: free ions, ion pairs, polarized complexes, and so on. Shifting this equilibrium to one or another side by varying polymerization conditions (temperature, solvent nature) allows influencing on kinetics of the process and structure of forming polymers.

Termination is a mandatory elementary stage of radical polymerization and usually proceeds through bimolecular reaction between two propagating macroradicals. Realization of the similar process in ionic polymerization is impossible due to Coulomb repulsion of reaction centers bearing the like-signed charges. Termination in ionic polymerization is realized with participation of molecules or particles which do not take part in propagation. In some cases, thorough removal of such particles from reaction media allows to realize "living" or "immortal" polymerization proceeding up to full consumption of monomer and capable of reinitiation after addition of a fresh feed of monomer.

Depending on the sign of the ion at the propagating end of macroradical cationic and anionic polymerization is distinguished.

(a) **Anionic polymerization**

In general, vinyl monomers capable of polymerizing by the anionic mechanism are those with substituents capable to stabilize carbanions [51]. Among monomers that tend to polymerize via anionic mechanism are ones bearing electron acceptor substituents at double bonds, for example 1,1-dicyanoethene, nitroethylene, acrylonitrile, methacrylonitrile, styrene, acrylates and methacrylates. Besides, anionic polymerization may be realized in case of alkylene oxides, lactones, lactams, and some carbonyl compounds such as aldehydes via cleavage of C=O bond. Disposition of vinyl and diene monomers toward anionic polymerization generally increases with the increase of electron-accepting properties of substituents. However, although ethylene does not have an anion-stabilizing substituent, it undergoes polymerization to high molecular weight polymer under suitable conditions with the use of the highly reactive initiator system comprising alkyllithium and tetramethylethylenediamine [52].

Catalysts of anionic polymerization are electron-donating compounds: alkali metals, its amides, and solutions in liquid ammonia and other solvents capable for electron solvation, organic compounds of alkali metals, and so on.

The general mechanism of anionic polymerization under action of alkali metal-based catalyst may be depicted by the following scheme:

Initiation:

$$RMt \rightleftharpoons R^{-}Mt^{+} \tag{2.34}$$

$$\overset{-}{R}\,\overset{+}{Mt} + H_2C{=}\underset{X}{\underset{|}{CH}} \longrightarrow R{-}H_2C{-}\underset{X}{\underset{|}{\overset{-}{CH}}}\,\overset{+}{Mt} \tag{2.35}$$

where X is -C_6H_5, -CN, –$CH{=}CH_2$, –COOR, etc., Mt is metal atom.

Propagation:

$$R{-}H_2C{-}\underset{X}{\underset{|}{\overset{-}{CH}}}\,\overset{+}{Mt} + H_2C{=}\underset{X}{\underset{|}{CH}} \longrightarrow R{-}H_2C{-}\underset{X}{\underset{|}{CH}}{-}H_2C{-}\underset{X}{\underset{|}{\overset{-}{CH}}}\,\overset{+}{Mt} \tag{2.36}$$

The length of polymer chain in anionic polymerization is determined by chain transfer reactions (via abstraction of proton from solvent or monomer by reaction center or by hydride transfer from propagating polymer chain to counterion or monomer reactions) or by isomerization of reaction center resulting in decrease of its reactivity. Chain transfer reaction on solvent proceeds during anionic polymerization of styrene catalyzed by solution of metal potassium in liquid ammonia:

$$\sim CH_2{-}\underset{Ph}{\underset{|}{\overset{-}{CH}}}\,\overset{+}{K} + NH_3 \longrightarrow \sim CH_2{-}\underset{Ph}{\underset{|}{CH_2}} + KNH_2 \tag{2.37}$$

In some cases, termination reactions may be avoided in anionic polymerization. It results in remaining of macroanions capable for propagation in reaction vessel even after full consumption of monomer. The number of such "living" chains is equal to the quantity of initially introduced catalyst with the deduction of chains reacted with impurities. In a simple case, polymerization kinetics is determined only by ratio by initiating (k_{in}) and propagating (k_p) constants. If $k_{in} \gg k_p$ following simple equations are true for polymerization rate and molecular weight:

$$V_p = k_p[\mathrm{M}][\mathrm{I}_0] \tag{2.38}$$

$$\overline{P_n} = nq\frac{[\mathrm{M}_0]}{[\mathrm{I}_0]} \tag{2.39}$$

where q is degree of monomer conversion, n number of reaction centers per one macromolecule, and $[\mathrm{M}]_0$ and $[\mathrm{I}]_0$ initial concentrations of monomer and initiator in polymerization media. In case of $k_{in} \gg k_p$, when all polymer chains start growing simultaneously, formation of polymers with very narrow molecular weight distribution ($M_w/M_n \leq 1.1$) may be achieved. The rate of anionic polymerization as well as cationic one significantly depends on solvent nature and usually increases with its dielectric constant.

(b) **Living anionic polymerization**

Anionic polymerization is the first chain polymerization, which could be conducted without the occurrence of chain termination and chain transfer. Such polymerization was given the name "living" by its discoverer M. Szwarc. The discovery made it possible to exert immense control on polymerization, molecular weight and its distribution, polymer end groups, and molecular architecture [13, 53]. Most of the foundational works on living polymerization were done using styrene, butadiene, isoprene, and methyl methacrylate as monomers.

In the first living polymerization described by Szwarc, styrene was polymerized at room temperature in THF. Sodium naphthalene was used as the initiator by virtue of its solubility and ability to initiate polymerization rapidly by electron transfer to monomer, which is important for achieving polymer of narrow molecular weight distribution (MWD) [13, 53, 54]. Although polymerizations of nonpolar monomers such as styrene, butadiene, and cyclopentadiene initiated by sodium naphthalene in ethereal solvents were discovered much earlier by Scott, the living nature of these polymerizations was not recognized then [55].

Hence, the polymerization must be performed using thoroughly purified monomers, solvents, and reagents in vessels from which air and moisture have been scrupulously excluded. However, purposeful termination ("killing") of the polymerizations with appropriate reagents leads to technologically useful end-functionalized polymers [54, 55].

Effect of additives: Polar additives such as ethers, tertiary amines, and sulfides decrease or even eliminate aggregation of organolithium compounds effecting increase in rates of initiation and propagation [56, 57]. These compounds interact specifically by coordinating with lithium. The tetramethylethylenediamine is very effective in this regard. It forms a monomeric chelated complex, which is soluble in hydrocarbon solvents in all proportions. The complex is a powerful initiator, which is attributed to not only its monomeric nature but also the ionic character of the Li–C bond being increased to that of the corresponding ion pair. Dissociation to free ions may also occur to some extent [13, 58, 59].

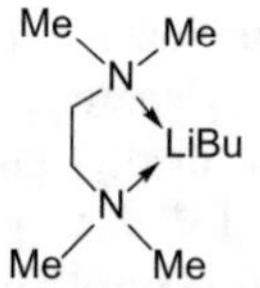

Me Me N LiBu N Me Me (2.40)

The effect of additives on initiation is much larger than on propagation, which may be due to the higher degree of aggregation of the organolithium initiator compared to that of the propagating chain end (vide supra). Thus, using anisole as the additive, to the extent of five to 40 times the equivalent concentration of *n*-BuLi, initiation of polymerization of styrene or dienes is completed in minutes at 30 °C, while propagation remains virtually unaffected. However, different additives affect the polymer microstructure and the chain end stability differently. Thus, whereas the *cis*-1,4 structure in polyisoprene is not formed at all in THF, even when used in small amounts, diphenyl ether and anisole do not affect the microstructure to any great degree, even when used undiluted [60]. On the other hand, the chain end stability is higher in anisole than in THF [61].

In contrast to the above nonionic Lewis bases, the ionic additives such as lithium alkoxides retard polymerization initiated by organolithium compounds. This is because they are aggregated more strongly than polystyryl- or polydienyllithium propagating species. Mixed complex formed between the propagating species and the more aggregated LiOR increases the state of aggregation of the former. As a result, the reactivity of the propagating species decreases and retardation of polymerization occurs [62].

Stereospecificity: Polymerization of isoprene initiated by lithium or alkyllithium initiators in bulk or in hydrocarbon solvents yields *cis*-1,4-polyisoprene with more than 93% *cis* content similar to natural rubber. Initiation by other alkali metals such as sodium and potassium gives *trans*-3,4- and *trans*-1,4-polyisoprenes [63, 64]. Polar compounds bring about changes in microstructure [60, 65]. For example, 1,4-addition does not occur in THF. When butadiene is used in place of isoprene, the microstructure is again largely 1,4 but of mixed *cis*- and *trans*-geometry when the initiator is lithium.

The all *cis*-1,4 structures of polyisoprene promoted by lithium or its alkyl derivatives may be explained based on the coordination of the diene with Li^+ ion in the chain end prior to incorporation into the chain.

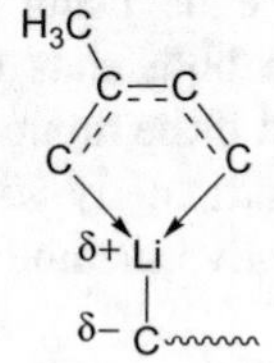

H_3C C C C C $\delta+$ Li $\delta-$ C (2.41)

Although butadiene can also coordinate with Li^+ ion, the *cis*-configuration in the last monomer unit is retained in polyisoprene due to the prevention of rotation by the steric hindrance exerted by the methyl group. A polar solvent such as THF, which strongly coordinates with Li^+ ion, precludes coordination with dienes resulting in loss of microstructure control. In contrast, polymerization of polar monomers in THF occurs stereospecifically. For instance, polymerization of MMA in THF yields *syndio*-rich living polymer [66, 67].

Synthesis of block copolymers by means of living polymerization: The direct method of synthesis of block copolymers involves sequential living polymerization of the concerned monomers in the order of increasing electronegativity. This is because the living polymer prepared from an electronegative monomer may not succeed in initiating the polymerization of an appreciably less electronegative one [52]. For instance, the synthesis of a diblock copolymer of styrene and MMA may be successfully done when the appreciably less electronegative styrene is polymerized first using a monofunctional initiator but not when MMA is polymerized first. However, the synthesis will not be clean in as much as the polystyryl anion or the ion pair is too high in nucleophilicity to be incapable of inducing side reaction in the polymerization of MMA. The problem is overcome by way of reducing nucleophilicity, for example, by adding a few drops of 1,1-diphenylethylene to the living polystyrene solution. This monomer does not homopolymerize but adds to polystyrene living end giving the desired result [68].

In contrast, when two monomers differ in electronegativity only to a small degree, for example, styrene and butadiene, or styrene and isoprene, their diblock copolymers can be synthesized irrespective of the sequence followed. However, polymerizing dienes first in a hydrocarbon solvent gives polystyrene block with broader MWD. This is due to the greater degree of association of polydienyllithiums making initiation of styrene polymerization slower than propagation. The shortcoming in this case can be rectified by adding some polar solvent before polymerization of styrene is undertaken. The polar solvent disintegrates the aggregates of both polydienes and polystyrene living ends forming monomeric species.

This strategy is followed in the synthesis of the thermoplastic elastomers, polystyrene-polybutadiene-*b*-polystyrene (SBS), or polystyrene-*b*-polyisoprene-*b*-polystyrene (SIS) in hydrocarbon solvents [69].

The ABA triblock copolymers may be synthesized also by first synthesizing the middle B-block with two living ends using a difunctional initiator and then extending the polymer chain from both ends using monomer A. However, this sequential method does not succeed if the monomer sequence does not observe the increasing electronegativity rule, particularly when the electronegativity difference is quite large. However, the alternative method described above can be used successfully in this case also.

Synthesis of star polymers: Three methods exist for the synthesis of star polymers during anionic polymerization [70]:

1. Initiation by a multifunctional initiator

$$\text{Mt}^{+}\ {}^{-}\text{(trifunctional initiator)} \xrightarrow{\text{monomer}} \text{(three-arm star with Mt}^{+}\text{ ends)} \qquad (2.42)$$

2. Termination by a multifunctional terminator

$$\text{X}_3\text{(trifunctional terminator)} + 3\,\text{Mt}^{+}\,{}^{-}\text{\~\~\~} \longrightarrow \text{(three-arm star)} + 3\,\text{Mt}^{+}\text{X}^{-} \qquad (2.43)$$

3. Polymerization of a difunctional monomer initiated by a living chain

$$\text{Mt}^{+}\,{}^{-}\text{\~\~\~} + \text{divinylbenzene} \longrightarrow \text{(star polymer with Mt}^{+}\text{ ends)} \qquad (2.44)$$

(c) **Cationic polymerization**

Cationic polymerization is a convenient tool for polymerization of vinyl and diene monomers bearing electron-donating substituents at double bond, for example isobutylene, α-methyl styrene, vinylalkyl esters, isoprene [71, 72].

The increase of donating ability of substituent results in increase of monomer tendency to participate in cationic polymerization. Some carbonyl compounds (e.g., formaldehyde), alkyne oxides, and other heterocyclic compounds also may participate in cationic polymerization. Cationic polymerization is catalyzed by electron-accepting compounds such as strong protonic acids (H_2SO_4, $HClO_4$, etc.) and Lewis acids (BF_3, $SnC1_4$, $TiCl_4$, $AlBr_3$, $FeCl_3$, etc.). Catalysis of cationic polymerization by Lewis acid usually requires a cocatalyst such as water, protonic acids, alcohols, alkyl halides, ethers forming complexes with catalyst [73, 74].

Initiation of cationic polymerization of isobutylene in the presence of BF_3 and traces of water as cocatalyst may be described by the following scheme:

$$BF_3 + H_2O \rightarrow H^+[BF_3OH]^- \qquad (2.45)$$

$$H^+[BF_3OH]^- + CH_2{=}C(CH_3)_2 \rightarrow (CH_3)_3C^+[BF_3OH]^- \qquad (2.46)$$

As a result of this process, an active carbonium cation and corresponding counterion are formed. In the media with low dielectric constant, generated ions form ion pairs which are rather stable. Propagation of polymerization chain processes via consecutive addition monomer units to cation:

$$(CH_3)_3C^+[BF_3OH]^- + CH_2{=}C(CH_3)_2 \rightarrow (CH_3)C{-}CH_2{-}\overset{+}{C}(CH_3)_2[BF_3OH]^- \quad (2.47)$$

Chain termination in cationic polymerization may be realized via various reactions. At the same time, many of these reactions that terminate the growth of a propagating chain do not, however, terminate the kinetic chain because a new propagating species is generated in the process. Contrary to radical polymerization, termination of cationic polymerization has the first order with respect to propagating centers. Decay of active centers may proceed by interaction of macrocation with counterion or by formation of covalent bond instead of ionic. The latter is observed during polymerization of styrene catalyzed by CF3COOH:

$$\sim CH_2{-}\overset{+}{C}H(Ph)\left[{}^{-}O{-}C({=}O){-}CF_3\right] \longrightarrow \sim CH_2{-}CH(Ph){-}O{-}C({=}O){-}CF_3 \quad (2.48)$$

Another mechanism of termination is transfer of ionic group from counterion to growing carbocation:

$$\sim CH_2{-}C(CH_3)_2^+[TiCl_4OH]^- \longrightarrow \sim CH_2{-}C(CH_3)_2OH + TiCl_4 \quad (2.49)$$

A crucial role in cationic polymerization belongs to chain transfer reactions as it determines the length of macromolecules formed during polymerization. Chain transfer to monomer involves transfer of a β-proton to monomer with the formation of novel propagating chain and macromolecule with terminal unsaturation.

$$\sim CH_2{-}C(CH_3)_2^+ + CH_2{=}C(CH_3)_2 \longrightarrow \sim CH_2{-}CH(CH_3)_2 + CH_2{=}C(CH_3){-}\overset{+}{C}H_2 \quad (2.50)$$

Chain transfer to counterion proceeds by transfer of β-proton to the counterion. This process is reversible to initiation and results in regeneration of initiator which can further interact with novel molecule of monomer. As in chain transfer to monomer, the polymer molecule formed has a terminal double bond. Solvent molecules also may participate in chain transfer reactions.

Almost every system for cationic polymerization consisted of monomer, catalyst, cocatalyst, and solvent is characterized by special features. That is why it is hard to propose unified kinetic scheme. For most systems, polymerization rate has the first order toward catalyst concentration, while molecular weight is independent on catalyst concentration.

In common, rates of elementary stages of polymerization may be expressed as:

Initiation:

$$V_{in} = k_{in}[I] \tag{2.51}$$

Propagation:

$$V_p = k_p[P^+][M] \tag{2.52}$$

Termination:

$$V_0 = k_0[P^+] \tag{2.53}$$

Chain transfer:

$$V_{tr} = k_{tr}[P^+][M] \tag{2.54}$$

where [I] is initiator concentration; [M] monomer concentration; and [P⁺] concentration of active centers.

In a steady-state conditions, when $V_{in} = V_0$, polymerization rate may be described as:

$$V_p = \frac{k_{in}k_p}{k_0}[M][I] \tag{2.55}$$

The average degree of polymerization is expressed as the propagation rate divided by the sum of termination and transfer rates:

$$\overline{P_n} = \frac{V_p}{V_0 + V_{tr}} \tag{2.56}$$

or

$$\frac{1}{\overline{P_n}} = \frac{k_0}{K_p[M]} + \frac{k_{tr}}{k_p} \tag{2.57}$$

Thus, polymerization rate depends on initiator concentration, while molecular weight is independent on it.

(d) **Living carbocationic polymerization**

The reaction sequences in living carbocationic polymerization involving reversible termination are represented by the following equations:

Initiation:

$$R{-}X + MtX_n \rightleftharpoons R^+ MtX_{n+1}^- \tag{2.58}$$

$$R^{+}MtX_{n+1}{}^{-} + M \rightarrow RM^{+}X_{n-1}{}^{-} \quad (2.59)$$

Propagation:

$$RM_{k}{}^{+}X_{n-1}{}^{-} + M \rightarrow RM_{k+1} + X_{n-1}{}^{-} \quad (2.60)$$

Reversible termination:

$$RM_{k+1}{}^{+}X_{n-1}{}^{-} \rightleftharpoons RM_{k+1} - X + MtX_{n} \quad (2.61)$$

In the scheme, RX is an organic halide initiator with a labile C–X bond and MtX_n is a relatively weak Lewis acid. The ionic species has been shown as ion pair since living cationic polymerization is best achieved with it rather than with free ion (vide infra) [75–77].

The following conditions apart from the absence of transfer and termination are required to be fulfilled in order to achieve polymer with low PDI:

(1) Initiation should not be slower than propagation.
(2) The deactivation equilibrium must be dynamic, and the exchange between active and dormant states must be much faster than propagation so that a chain undergoes a large number of activation–deactivation cycles during the whole course of polymerization.
(3) In addition, adventitious initiation notoriously associated with carbocationic polymerization should be absent or reduced to an insignificantly low level.

The living nature of carbocationic polymerizations of several monomers was ascertained by various research groups from the following results:

(1) Monomer disappearance follows first-order kinetics.
(2) M_n increases linearly with conversion.
(3) Initiator efficiency is close to unity.
(4) The PDI is low ca., ~ 1.1.
(5) Terminal unsaturation in polymer is absent.
(6) Block copolymers are obtained using the living polymers as macroinitiators in the polymerizations of suitable monomers.

In some cases during homo- and copolymerization via ionic mechanism, a formation of complex between active center and monomer molecule precedes its incorporation into polymer chain. The lifetime of such complex may exceed a lifetime of transition state in common chain reactions (10^{-13} s) indicating its high stability. In such conditions, a case of a so-called coordination ion polymerization is observed. Solvents have a high influence of stability of such complexes. This fact causes a dependence of composition of copolymers obtained by ionic polymerization on solvent nature. A coordination ion polymerization mechanism determines stereoregularity during polymerization.

2.5 Stereoregulation in Radical and Ionic Polymerization

In case of polymerization of nonsymmetric monomers of CH_2=CHX or CH_2=CXY types, two neighboring units may be connected as in "head-to-tail" (a) so in a "head-to-head" (b) manner:

$$\sim CH_2-\overset{X}{\underset{H}{\overset{*}{C}}}-CH_2-\overset{X}{\underset{H}{\overset{*}{C}}}\sim \quad \text{a} \qquad \sim CH_2-\overset{X}{\underset{H}{\overset{*}{C}}}-\overset{X}{\underset{H}{\overset{*}{C}}}-CH_2\sim \quad \text{b} \tag{2.62}$$

"Head-to-head" addition in case of most vinyl monomers is connected with high activation energy barriers and thus may be excluded from consideration. Polymer chain constructed from monomer units connected in a "head-to-tail" order has a chiral carbon atom in every monomer in it:

$$R_1-\overset{X}{\underset{H}{\overset{*}{C}}}-R_2 \tag{2.63}$$

In case of similar configuration of all chiral atoms, a polymer chain is called isotactic (i). A syndiotactic (s) structure is characterized by strict alternation of chiral atom configurations in molecule. If configuration of chiral atoms is random, polymer structure is called atactic. The scheme below describes isotactic, syndiotactic, and atactic polymers drawn in the Fischer projections (Fig. 2.3).

$$\begin{array}{ccccccc} H & X & H & X & H & X & H \\ \hline H & Y & H & Y & H & Y & H \end{array} \quad \text{a} \qquad \begin{array}{ccccccc} H & X & H & Y & H & X & H \\ \hline H & Y & H & X & H & Y & H \end{array} \quad \text{b} \qquad \begin{array}{ccccccc} H & X & H & X & H & Y & H \\ \hline H & Y & H & Y & H & X & H \end{array} \quad \text{c} \tag{2.64}$$

Configuration of monomer unit in propagating polymer chain in radical polymerization is determined only after addition of the following monomer unit. It is concerned with possible rotation of terminal carbon atom-bearing unpaired electron around terminal carbon–carbon bond.

Formation of syndiotactic sequences usually has lower activation energy in comparison with isotactic ones. Thus, decrease of polymerization temperature tends

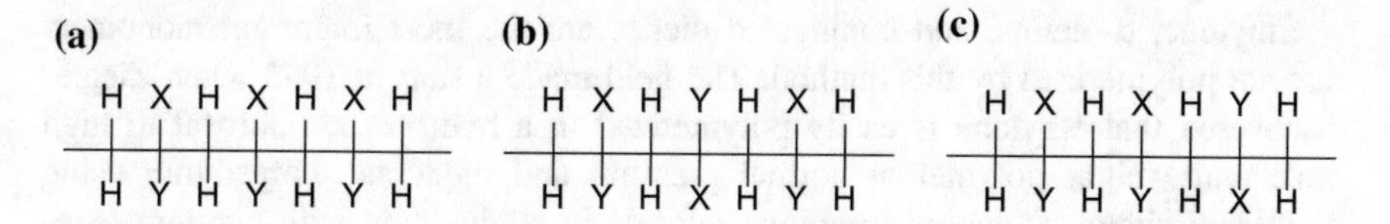

Fig. 2.3 Fischer projections for polymer chain with different structures: **a** isotactic, **b** syndiotactic, **c** atactic

to increase syndiotacticity. So, syndiotactic sequences are observed for about 80% of units in poly(methyl methacrylate) obtained at 80 °C. Decrease of polymerization temperature down to −70 °C results in increase of part of syndiotactic units up to 100%. The same tendency for increase of syndiotactic fragments on temperature decrease is observed in radical polymerization of other vinyl monomers. At the same time, the mentioned difference in activating energies for iso- and syndiotactic additions is rather small. According to this fact and rather high temperatures used for radical polymerization, the polymers formed mostly have no stereoregularity.

Ionic polymerization is characterized by higher stereoregularity in comparison with radical one. It governs not only by interactions between side substituents of propagating chain and monomer but by participation in chain propagation of other species, for example counterion. In case if ionic pair acts as an active center in polymerization process, counterion is included in formation of transition state forming during propagation. It may influence fixation of spatial configuration of the end of propagating chain. In some cases, the influence of counterion comes only to spatial effects. It may be considered as additional substituent at propagating end of macromolecule. For example, cationic polymerization of vinyl isobutyl ether in the presence of $\mathbf{BF_{3*}H_2O}$ (with $[BF_{3*}OH]^-$ as counterion) at −70°C leads to atactic polymer, while using $\mathbf{BF_{3*}(C_2H_5)_2O}$ catalyst (with $\mathbf{[BF_{3*}OC_2H_5]^-}$ as counterion) results in isotactic polymer. In this case the increase of the volume of counterion significantly increases the control over sterioregularity of the forming polymer.

For most catalytic systems, conducting stereospecific polymerization addition of monomer to propagating polymer chain is preceded by formation of complex between active center and monomer molecule. Coordination of monomer molecule in complex determines stereospecific cleavage of double bond leading to specific configuration of the next monomer unit. Such mechanism of macromolecule formation is called coordinational–ionic.

2.6 Coordination Polymerization

Coordination polymerization is a process of polymer synthesis during which the monomer and the active center are coordinated to the polymerization catalyst prior to the incorporation of the monomer in the polymer chain. The polymerization exhibits various degrees of stereochemical control on the structure of the polymer depending on the nature of the catalyst and reaction conditions.

Ethylene, α-olefins, and conjugated dienes are the most important monomers that are polymerized by this method. The field made a start in 1953 when Ziegler discovered that ethylene is easily polymerized in a hydrocarbon solvent to high molecular weight polymer at normal pressure and moderate temperature using triethylaluminum or diethylaluminum chloride in conjunction with titanium tetrachloride as catalysts [78, 79]. In the following year, Natta discovered stereochemical control exerted by the catalysts on the structure of polypropylene. The discoveries not only gave a new dimension to polymer science but also revolutionized the polyolefin industry. Ziegler and Natta were honored with a Nobel Prize in 1963, and the catalysts were named after them.

The first-generation Ziegler–Natta catalysts are comprised of two components:

(1) Transition metal compound of Group IV–VIII, which may be a halide, a subhalide, an oxyhalide, an alkoxide, a β-diketonate, a cyclopentadienyl dihalide, and so on;
(2) A Group I–IV base metal alkyl or hydride. The transition metal compound is essentially the precatalyst, which is alkylated, and thus activated, by the base metal alkyl to form the real catalyst. In the literature, however, the two-component mixture is usually referred to as the catalyst or the catalyst system, although sometimes the base metal alkyl is referred to as the cocatalyst and the transition metal compound as the catalyst. The higher generations Ti-based catalysts supported on activated magnesium chloride contain a third and a fourth component, which are often Lewis bases, added to increase the activity and the stereospecificity. These additives also influence the molecular weight of the polymer to be formed. The first catalysts discovered, as already introduced, are $AlEt_3$ + $TiCl_4$ and $AlEt_2Cl$ + $TiCl_4$. The two components of the catalysts on mixing together in a hydrocarbon solvent react to yield the real catalyst, which is insoluble in the medium leading to heterogeneous catalysis [78]. The active sites of the catalysts are not identical. This was immediately evident when the Natta group obtained a mixture of both crystalline (*isotactic*) and amorphous (*atactic*) polypropylene in the polymerization of propylene [80].

The real catalyst from triethylaluminum and titanium tetrachloride is formed in reactions (2.28)–(2.31). At temperatures above ca., −30 °C and at low Al-to-Ti ratios, $TiCl_4$ is reduced to $TiCl_3$, which precipitates as crystalline solid (indicated by the subscript "s"), and $AlEt_3$ is converted to $AlEt_2Cl$ liberating an ethyl radical. The latter undergoes disproportionation and coupling reactions forming ethylene and butane, respectively. Further reaction of $TiCl_3$ crystals with $AlEt_2Cl$ occurs resulting in the alkylation of Ti on the crystal surface (reaction 2.36). The alkylated $TiCl_3$ crystals act as the real catalysts. In fact, mixtures of separately prepared crystalline $TiCl_3$ and $AlEt_2Cl$ make very effective catalysts. Polymerization takes place by insertion of monomer into Ti–C bond.

$$TiCl_4 + Al(C_2H_5)_3 \rightarrow TiCl_{3(s)} + Al(C_2H_5)_2Cl + C_2H_5^{\cdot} \quad (2.65)$$

$$TiCl_4 + Al(C_2H_5)_2Cl \rightarrow TiCl_{3(s)} + Al(C_2H_5)Cl_2 + C_2H_5^{\cdot} \quad (2.66)$$

$$2\,C_2H_5^{\bullet} \rightarrow C_4H_{10} \quad ; \quad 2\,C_2H_5^{\bullet} \rightarrow C_2H_4 + C_2H_6 \tag{2.67}$$

$$TiCl_{3(s)} + Al(C_2H_5)_2Cl \rightarrow Ti(C_2H_5)Cl_{2(s)} + Al(C_2H_5)Cl_2 \tag{2.68}$$

(a) **Mechanism of coordination polymerization**

Natta proposed a coordinated anionic mechanism, which involves coordination of monomer to transition metal (as a rule titanium or vanadium) in an active site of the catalyst followed by insertion of the coordinated monomer into the highly polarized transition metal–carbon bond, which is formed in situ through the alkylation of the transition metal in the precatalyst by the base metal alkyl cocatalyst, as discussed earlier [81]. In support of the perceived insertion into the Mt–C bond rather than into the base metal–carbon bond, Mb–C, it may be cited that some organometallic catalysts containing only titanium as metal atom also exhibit isospecificity [82, 83].

However, insertion may occur in either of the two regiochemical modes shown in Eq. 2.67 with propylene monomer. The mode of insertion depends on the nature of the transition metal as well as that of the carbon, i.e., *primary* or *secondary*, in the Mt–C bond. Insertion of propylene into titanium–carbon bond occurs overwhelmingly in the 1,2 mode (95–99% depending on the nature of the catalyst) [84, 85]. The regiospecificity of the *atactic* polymer is only little lower than that of the *isotactic* polymer. In contrast, insertion into vanadium–carbon bond shows high preference for the 2,1 mode, when the carbon is *secondary*, and only a low preference for the 1,2 mode when the carbon is *primary* [86–88]. As a result, long sequences of 2,1 insertions alternate with short sequences of 1,2 insertions. Overall, the 2,1 insertions comprise ca., 85% of the total. It has also been established that 2,1 insertions into vanadium–carbon bond in the soluble vanadium-based catalysts are syndiotactic, whereas 1,2 insertions are *atactic* [88]. Thus, the regioblocks of *syndiotactic* propylene are also stereoblocks comprising long *syndiotactic* blocks interposed by short *atactic* blocks.

$$\mathrm{Mt{-}R} + \overset{1}{C}H_2{=}\overset{2}{C}H{-}\overset{3}{C}H_3 \rightarrow \mathrm{Mt{-}CH_2{-}\overset{*}{C}H(CH_3){-}R} \;\; \text{1, 2 insertion} \quad ; \quad \rightarrow \mathrm{Mt{-}\overset{*}{C}H(CH_3){-}CH_2{-}R} \;\; \text{2, 1 insertion} \tag{2.69}$$

The 2,1 insertions yield tail-to-head linkages between monomers (the unsubstituted end of the monomer being tail and the substituted end being head), whereas 1,2 insertions yield head-to-tail linkages. A regioblock structure is essentially a multiblock of head-to-tail (H–T) and tail-to-head (T–H) polymers.

The absolute configuration of the chiral carbon in each newly added monomer unit (Eq. 2.70) depends on not only the regiochemistry but also the stereochemistry of monomer insertion. As discussed above, the monomer insertion into the Ti–C bond is regiochemically almost completely 1, 2. The stereochemistry of insertion is always *cis* irrespective of the type of propagation, *isospecific* or *syndiospecific* [89, 90]. However, it depends on the prochiral face of the olefin involved in the reaction. The difference in the stereochemistry of *cis* 1,2-insertion of α-olefin into the Ti–C bond is shown in Eq. 2.70 for the involvement of the two different prochiral faces of the olefin in the reaction.

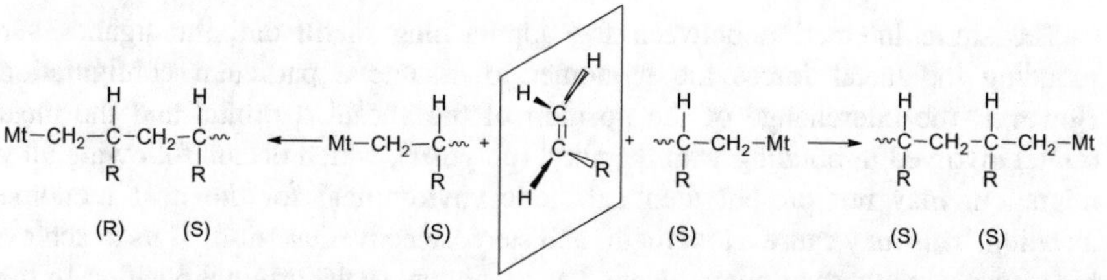

Mt–CH$_2$–C(H)(R)–CH$_2$–C(H)(R)~ ⟵ Mt–CH$_2$–C(H)(R)~ + H$_2$C=C(H)R + ~C(H)(R)–CH$_2$–Mt ⟶ ~C(H)(R)–CH$_2$–C(H)(R)–CH$_2$–Mt

(R) (S) (S) (S) (S) (S)

(2.70)

Thus, for isospecific propagation, the active center must discriminate between the two prochiral faces of the α-olefin so that the same prochiral face is attacked in each insertion step [91]. For this to happen, it must have center(s) of chirality. Looking at the structure **5** of the active center, the β carbon atom (with respect to the metal) in the chain head is chiral. Besides, the metal atom itself can be a center of chirality. However, it should maintain its absolute configuration during the insertion process. The latter can be achieved by the bound to the solid catalyst surface [84].

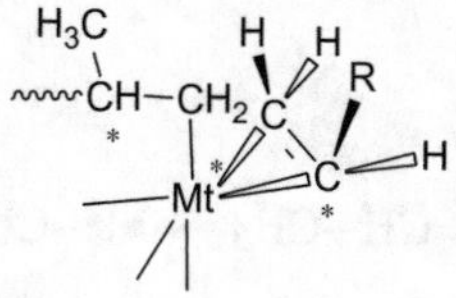

~CH(CH$_3$)–CH$_2$–Mt* ··· H$_2$C=C*(H)R

(2.71)

Direct evidence demonstrating the chirality of the transition metal atoms in isospecific catalysts was obtained from the observation that the heterogeneous isospecific catalyst $TiCl_3/AlR_3$ polymerizes a *racemic* α-olefin to the corresponding *racemic* polymer. Thus, the chiral metal, overall, is racemic [91]. In contrast, the homogeneous aspecific catalyst $Ti(CH_2C_6H_5)_4/Al(CH_2C_6H_5)_3$ produces a copolymer of the two enantiomers of the *racemic* monomer, which cannot be resolved by elution chromatography using an optically active stationary phase. On the other hand, there are evidences which suggest that the chirality of the β-carbon atom at the chain head is not essential for isospecific propagation.

The mechanism of insertion was proposed by Cossee and Arlman [92]. The mechanism (Eq. 2.72) proposes a concerted four-center reaction involving titanium and carbon atoms of the Ti–C bond in the titanium alkyl (polymer) and the two π-bonded carbon atoms of the olefin. A four-membered ring transition state is formed, which is followed by migration of the alkyl (polymer) to one end of the olefin. The π-bonding between titanium and the olefin effects weakening of the labile Ti–C bond, which facilitates the migration of the alkyl (polymer) group.

(2.72)

The steric interaction between the approaching olefin and the ligands surrounding the metal forces the monomer to assume a particular configuration. However, the interchange of the position of the vacant d orbital and the metal orbital involved in bonding with the alkyl (polymer), which occurs following alkyl migration, may not present identical steric environment for the next monomer insertion. This may cause a less regio- and stereoselective insertion. Thus to achieve high isospecificity, the polymer chain has to migrate to the original position so that the monomer occupies the same vacant site before each insertion step.

Chain transfer and regulation of molecular weight. Two types of thermal cleavage may occur in olefin, for example propylene polymerization. The β-hydride elimination shown in Eq. (2.73) may proceed without assistance of the monomer [93]. However, the metal hydride produced may subsequently react with the monomer. The two steps together represent monomer transfer. However, reaction (2.75) is insignificant at low temperature with Ti-based catalysts. Reaction (2.75) also represents a monomer transfer, although the transferred monomer retains the double bond.

$$\mathrm{Mt{-}CH_2{-}CH(CH_3){\sim\sim}} \xrightarrow{\text{below } 60\,^\circ\mathrm{C}} \mathrm{Mt{-}H + CH_2{=}C(CH_3){\sim\sim}} \quad (2.73)$$

$$\mathrm{Mt{-}H + CH_2{=}CH{-}CH_3 \rightarrow Mt{-}CH_2{-}CH_2{-}CH_3} \quad (2.74)$$

$$\mathrm{Mt{-}CH_2{-}CH(CH_3){\sim\sim} + CH_2{=}CH{-}CH_3} \xrightarrow{>80\,^\circ\mathrm{C}} \mathrm{Mt{-}CH_2{-}CH{=}CH_2 + {\sim\sim}CH(CH_3){-}CH_3} \quad (2.75)$$

Another type of transfer is alkyl transfer:

$$\mathrm{Mt{\sim\sim} + Mb{-}R \longrightarrow Mb{\sim\sim} + Mt{-}R} \quad (2.76)$$

where Mt is the transition metal and Mb is the base metal.

Hydrogen is also used as a transfer agent to regulate the molecular weight [94]. The reactions involved may be represented as follows:

$$\text{Mt}\sim\sim + H_2 \longrightarrow \text{H}\sim\sim + \text{Mt–H} \quad (2.77)$$

$$\text{Mt–H} + CH_2{=}CH{-}CH_3 \rightarrow \text{Mt}{-}CH_2{-}CH_2{-}CH_3 \quad (2.78)$$

(b) **Metallocene catalysts**

The original metallocene Ziegler–Natta catalysts are comprised of two components —a Group IV metallocene, e.g., zirconocene or titanocene being one, and methylaluminoxane (MAO) being the other. The metallocene is actually the precatalyst, which is activated by MAO to form the real catalyst. These are soluble catalysts, and the active sites in the catalytic species are all of the same type. Hence, they are referred to as single-site catalysts. These catalysts produce polyolefins with lower polydispersities and more uniform distribution of comonomers in copolymers of ethylene and α-olefins than are obtained using Ziegler–Natta catalysts. Besides, they are much more active. The polymers exhibit better film clarity and tensile strength, and contain lower amounts of oligomers. Metallocene catalysts were discovered by Kaminsky and coworkers about 25 years after the discovery of the classical Ziegler catalysts [95]. MAO is prepared by the controlled hydrolysis of trimethylaluminum [96, 97]. It has the basic unit.

$$(H_3C)_2Al{-}O{-}Al(CH_3){-}O{-}Al(CH_3){-}O{-}Al(CH_3)_2 \quad (2.79)$$

These units associate to satisfy the unsaturated aluminum atoms through coordination in the best way possible. Three to four such units form a cage structure, which is believed to be its active form [98]. In analogy to methyl siloxane, the compound is so named. The corresponding ethyl aluminoxane is not very active. However, MAO is required to be used in very high excess, e.g., MAO: metallocene = 5000:1 (mol/mol).

Zirconocene/MAO catalysts exhibit very high activity. For example, biscyclopentadienyl–zirconium dichloride and MAO reportedly produced nearly 40 metric tons of polyethylene (molecular weight 78,000) per g zirconium in 1 h at 95 °C and 8 bar ethylene pressure. This sort of activity is ten to hundred times larger than that observed with the classical Ziegler catalysts. Additionally, the molecular weight distribution is narrower, $M_w/M_n \approx 2$. Oligomers are formed only in traces.

Metallocene catalysts also proved to be much more active than the vanadium-based Ziegler–Natta catalysts in the copolymerization of ethylene and cycloolefins, such as cyclobutene, cyclopentene, and norbornene, without ring opening of the cyclic olefins taking place [99].

For *isotactic* propagation of polypropylene chain, a chiral metallocene is required to be used. However, the mobility of the ligands bonded around the metal center must be either prevented or rendered slower than propylene insertion. This is achieved in *ansa*-metallocenes. In such a metallocene, the two cyclopentadiene ligands sandwiching the metal are substituted to make them rigid. Furthermore, they are linked with a $-CH_2-CH_2-$ bridge. The cyclopentadiene ligands are thus, so to say, strapped (*ansa*) around the metal. The *ansa*-zirconocene, ethyl(tetrahydroindenyl)$_2$zirconium dichloride, [Et(THind)$_2$ZrCl$_2$], has three stereoisomers, *R*, *S*, and *meso*. The *meso* isomer produces the *atactic* polypropylene, whereas the *R* or *S* isomer produces the *isotactic* polymer. Suitable substituents in the indenyl rings and bridges linking the latter have led to catalysts, which produce polypropylenes with very high *isotacticity* ca., 99% [97].

For *syndiotactic* propagation, the catalyst should have Cs symmetry [100]. It has a bent sandwich structure, which provides two different bonding positions for the inserting monomer. *Syndiotacticity* is believed to originate from active site isomerization with each monomer addition.

Late transition metal catalysts hold promise of producing polyethylene or poly (α-olefin)s with novel microstructure leading to novel properties and applications. Besides, due to the low oxophilicity they have much greater functional group tolerance than early transition metal ones have and, accordingly, are suitable for the copolymerization of ethylene or α-olefins and polar comonomers.

α-Diimine complexes of Pd(II) and Ni(II) and pyridyl bisimine complexes of cobalt and iron proved effective catalysts in the polymerization of ethylene, α-olefins, and cyclic olefins or copolymerization of olefins and polar monomers [101–104]. The catalyst activity may be tuned by appropriately changing the substituents.

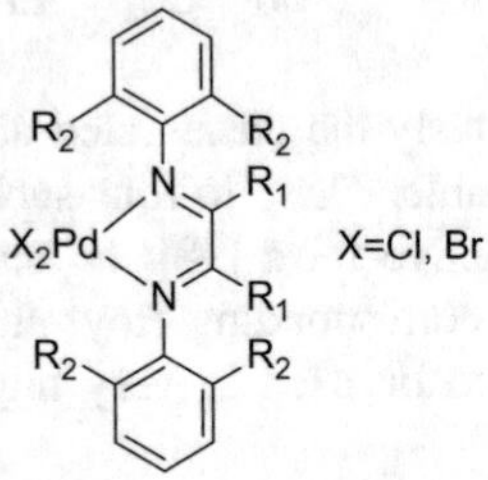

X_2Pd ... X=Cl, Br (2.80)

The pyridyl bisimine complexes of cobalt and iron [103–105] are also a promising catalysts for such reactions. One of the most active iron-based catalysts has bulky aryl substituents in the imine groups as shown in complex.

By using different R and R′ groups, the activity of the catalyst and the molecular weight of the polymer can be further tuned. A five-coordinate pseudo-square pyramidal conformation for the complex has been suggested from crystallographic studies. The aryl groups on the imino nitrogens stand nearly perpendicular to the ligand coordination plane [102]. Bis-chelate formation is prevented by the bulky substituents on the imino nitrogens resulting in high activity for ethylene polymerization [103]. The polyethylene obtained is strictly linear and of high molecular weight in contrast to the highly branched polyethylene obtained with Pd diimine catalysts discussed above. The late metal catalysts, however, have not yet found commercial use in polyolefin production. Many of these catalysts initially exhibit very high activity, which goes on decreasing during the course of polymerization.

(c) **About living polymerization of alkenes**

Natta and coworkers provided the first indication of living polymerization of propylene conducted at −78 °C using the soluble catalyst VCl_4 + $AlEt_2Cl$ [106]. The molecular weight of the *syndio*-rich polypropylene increased linearly with time over a period of 25 h. Later, Doi et al. found the molecular weight polydispersity to be between 1.4 and 1.9, which is rather high for a living polymer. They, however, succeeded to obtain *syndio*-rich polypropylene of low PDI (PDI = 1.07–1.18) replacing VCl_4 with V(acetylacetonate)$_3$ [107, 108]. The molecular weight of the polymer increased linearly with time over 15 h. The living character is appreciably lost by working at temperatures, which are even slightly greater than −65 °C. However, only one polymer chain forms per twenty-five vanadium atoms. The situation improves to yield one polymer chain per vanadium atom by replacing acetylacetonato with 2-methyl-1,3-butanedionato ligand.

Living polymerization by metallocene catalysts has also been achieved. Monomer transfer is prevented by resorting to low reaction temperature, while alkyl transfer is eliminated by using boron-based activators [109]. Various other catalysts including the late transition metal ones and activators have yielded living polymerizations. However, because of the low turnover (moles of monomer polymerized per mole of catalyst) the living coordination polymerization is yet to be adopted by the industry.

2.7 Ring-Opening Polymerization

A wide variety of cyclic compounds each containing at least one heteroatom or an unsaturation center in the ring undergoes ring-opening polymerization (ROP), as shown schematically in reaction

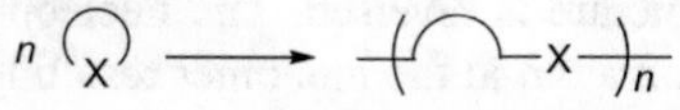

$$n \; \text{(cyclic-X)} \longrightarrow \left[\frown \text{—X—} \right]_n \qquad (2.81)$$

where X is a heteroatom or a double bond.

The mechanism of ring opening, however, is different for the two groups of monomers [110]. Whereas a heteroatom–carbon bond is opened in the heterocycles through a nucleophilic reaction, the double bond is opened through an olefin metathesis reaction in the cycloalkenes. The polymerization of the latter group of monomers is specifically referred to as ring-opening metathesis polymerization (ROMP). The difference in mechanism notwithstanding, many of the polymerizations of both groups of monomers are living in nature exhibiting chain growth without termination and irreversible chain transfer.

The polymer produced by ROP structurally resembles a condensation polymer. Indeed, some of these polymers are also produced by step polymerization (nylon 6 and silicone polymers being typical examples). However, ROP has several superior features. Step polymerization yields polymer of only moderate molecular weight and that too only at very high extents of reaction. For example, the DPn reaches only 200 at an inordinately high conversion of 99.5% that requires a long time to reach. Besides, an extremely high degree of monomer purity, accurate stoichiometric balance between the monomers, prevention of loss of monomers, and the absence of side reactions are absolute necessities. In contrast, ROP easily yields high molecular weight polymer at appreciably lower conversion and lesser time, and under fewer demanding conditions. Besides, many of these polymerizations, as pointed out above, are living and, accordingly, have the attractive features that go with it. Several polymers produced by ROP have found important biomedical applications besides various other applications: poly(ethylene oxide) and its block copolymers, polyglycolide, polylactides, polyanhydrides and their block copolymers, homopolypeptides and block copolypeptides, silicones, and polyphosphazenes being some examples.

Polymerization of heterocyclic monomers typically takes place anionically and/or cationically with propagating species containing charged heteroatom, e.g., alkoxide, carboxylate, sulfide, silanoate, in anionic ROP, and cyclic *tertiary* oxonium, sulfonium, iminium, siloxonium, and some other heteroatomic monomers, in cationic ROP. These anionic and cationic species are much less reactive than carbanionic and carbenium ionic species, respectively. The mechanism of polymerization involves nucleophilic substitution. In cationic ROP, the heteroatom in monomer acts as a nucleophile. It attacks an α-carbon atom of the propagating cationic species resulting in chain extension with ring opening

$$\left(\begin{matrix}CH_2\\ CH_2\end{matrix}\right)\ddot{X} + \sim\overset{+}{X}\left(\begin{matrix}CH_2\\ CH_2\end{matrix}\right) \longrightarrow \sim X{-}CH_2 \quad H_2C{-}\overset{+}{X}\left(\begin{matrix}CH_2\\ CH_2\end{matrix}\right) \tag{2.82}$$

X is a heteroatom.

In anionic ROP, the picture is reversed. The nucleophilic propagating anionic species attacks an α-carbon atom in the monomer resulting in chain extension with ring opening

$$\sim X^- + X\left(\begin{matrix}CH_2\\ CH_2\end{matrix}\right) \longrightarrow \sim X{-}CH_2 \quad H_2C{-}\bar{X} \tag{2.83}$$

(a) **Cyclic olefins polymerization via metathesis mechanism**

Under the influence of suitable transition metal catalysts, cyclic olefins undergo ring-opening metathesis polymerization (ROMP), which involves cleavage of carbon–carbon double bond and intermolecular reassembly of the opened fragments forming a linear polymer with the unsaturation now linking the opened rings [111, 112]:

$$n \; \text{cycloalkene} \longrightarrow \left[\; \right]_n \qquad (2.84)$$

Polymerization of a cycloalkene, e.g., cyclobutene, cyclopentene, cycloheptene, or cyclooctene, produces the corresponding polyalkenamer. Cyclohexene does not polymerize for unfavorable thermodynamics of polymerization.

Early catalysts constituted of transition metal compounds in their high oxidation states such as $MoCl_5$ and WCl_6 used in combination with strong Lewis acids cocatalysts, aluminum alkyls/alkyl halides [113, 114]. However, these catalysts are ill-defined since the true catalysts formed, in situ, the alkylidene complexes (vide infra), decompose in the course of polymerization under the conditions used. Nevertheless, some of these catalysts found applications in the production of some commercially important polymers, notably polynorbornene and polyoctenamer.

$$n \; \text{cyclooctene} \longrightarrow \left[CH{-}(CH_2)_6{-}HC \right]_n \qquad (2.85)$$

The well-defined soluble initiators that give rise to living ROMP are mostly the Grubbs initiators, which are metallacycles and alkylidene complexes of Ti and Ru, respectively, and the Schrock initiators, which are alkylidene complexes of Ta, Mo, or W.

Mechanism of ROMP may be described by the following stages:

Initiation:

$$L_nMt{=}R + \text{cycloalkene} \longrightarrow L_nMt{=}R \cdots \text{cycloalkene} \longrightarrow L_nMt\text{–metallacyclobutane–}R \longrightarrow L_nMt{=}\text{ring-opened}{=}R \qquad (2.86)$$

Propagation:

$$L_nMt{=}\text{ring-opened}{=}R + (n\text{-}1)\,\text{cycloalkene} \rightleftharpoons L_nMt{=}\left[\; \right]_n{=}R \qquad (2.87)$$

Initiation involves coordination of the cyclic olefin monomer to the transition metal alkylidene (directly supplied or formed in situ from the initiator) and subsequent formation of a four-membered metallacyclobutane intermediate by [2 + 2]-cycloaddition. The latter undergoes retro [2 + 2] cleavage forming a new metal alkylidene linked to one end of the ring-opened monomer. Propagation involves repetition of the above steps on the metal alkylidene chain end.

The more active metal alkylidene complex initiators for ROMP are complexes of W, Ru, Mo, and some other metals. There exists great scope of tuning the initiators by the appropriate choice of ligands to effect living ROMP. Thus, the activity of the highly active tantalum-based initiator is decreased by replacing phenoxide ligands with bulky and electron-rich diisopropylphenoxide ligands [115]. Similarly, the activity of the W- and Mo-based initiators can be modulated by modifying the alkoxide ligands some of which are shown in the figure [116]. Electron-withdrawing ligands increase the activity. However, the Ru-based initiators exhibit the opposite behavior [117].

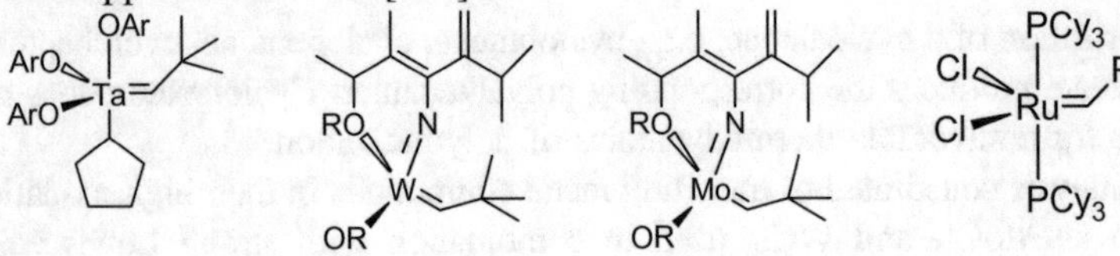

(2.88)

(b) **Living character of ROMP**

Ring-opening metathesis polymerizations effected by suitable Grubbs or Schrock initiators exhibit living character. They yield polymer of predicted molecular weight and low polydispersity (PDI < 2). The molecular weight increases linearly with conversion. The polymer formed may be chain-extended with other monomers of higher ring strain to yield block copolymers. The polymer can be "killed," for example, by Wittig-type reaction using aldehydes or ketones [117]. At the same time, ruthenium alkylidene-ended polymer is not attacked in these conditions. It may, however, be "killed" by alkyl vinyl ethers.

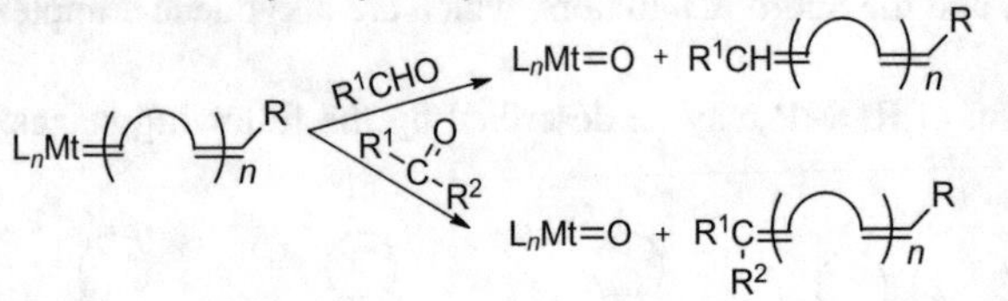

(2.89)

The catalyst's tolerance to the functional groups of the monomer is very important. In general, functional group tolerance of Ti- and Ta-based initiators is low, whereas that of W-, Mo-, or Ru-based initiators is high [117]. The Ru-based ones are the best, and the Mo-based initiators are better than the W-based ones. Both the Mo- and W-based initiators are, however, not tolerant to aldehydes and ketones. As a result, the living polymers prepared by using them may be "killed," for example, by benzaldehyde.

The low oxophilicity of Ru makes the Ru-based initiators extraordinarily tolerant toward functional groups and stable in protic solvents. Thus, such initiators are capable of polymerizing livingly a wide variety of functionalized norbornenes and cyclobutenes with such functional groups as hydroxyl, amino, amido, ester, and keto. Also, importantly, the polymerizations can be carried out in protic media. Even, water-soluble Ru-based initiators have been developed by incorporating charged groups in the phosphine ligands. These initiators are capable of polymerizing water-soluble norbornene derivatives at moderate temperatures ca., 45 °C in the presence of a protonic acid.

(c) **Polymerization of bridged cyclic monomers**

The simplest bridged cyclic monomers are the bicyclic monomers, which have three atoms common to the two rings as in norbornene (bicyclo [2.2.1] hept-2-ene) and

7-oxabicyclo[2.2.1] heptane. In these monomers, there are three bridges joining the two bridgeheads. Of these three, the one containing the heteroatom or the C=C bond opens for polymerization. Thus, norbornene polymerizes to form linear polymer with a cyclopentane ring in the repeating unit [118], whereas 7-oxabicyclo [2.2.1] heptane polymerizes to give linear polymer with a cyclohexane ring in the repeating unit [119].

(2.90)

(2.91)

2.8 Multimode Polymerization

The various methods of controlled/living polymerization described upper represent a powerful tool for producing macrostructures with various topologies including block, star, and graft copolymers through the sequential monomer addition. At the same time, each of described methods is limited by a certain scope of monomers capable to undergo polymerization with chosen system. Multimode polymerization proves to be a promising way to overcome the limitations of each method on synthesizing different block copolymers by using a single type of propagating species. In recent years, numerous block copolymer structures with both linear and branched macromolecular architectures were obtained by the combination of uncontrolled and/or controlled polymerization reactions. There are two general ways of multimode polymerization. The first one is based on using heterofunctional initiators, while the second implies transformation of end group in primary obtained macroinitiator to start polymerization via another mechanism.

(a) **Multimode polymerization using heterofunctional initiators**

This process is based on the use as initiators molecules containing two or more different initiation sites (A and B) that are capable of initiating concurrent polymerization mechanisms independently and selectively [120]. This technique allows to combine mechanistically incompatible monomers in one macromolecule without the necessity of intermediate transformation increasing selectivity. A requirement for effective realization of this mechanism is the high stability of each initiating center in polymerization conditions. The synthesis of block copolymers via heterofunctional initiators may be conducted as in sequential reaction steps, so in a one-pot reaction in case of controllable kinetics and the absence of side reactions.

sequential reactions

$$A\sim\sim B \xrightarrow{+M_A} P_A\sim\sim B \xrightarrow{+M_B} P_A\sim\sim P_B$$

$+M_A$ $+M_B$

one-step reaction

(2.92)

The significant breakthrough in the field of multimode polymerization is caused by intensive development of controlled polymerization techniques. In spite of this fact, pioneer works in this field appeared 20 years earlier. The use of bi- and polyfunctional radical initiators characterized by the simultaneous presence of at least two labile functional groups capable to generate active radicals at different conditions allowed obtaining polymers with ultrahigh molecular weight [121]. In case of sequential decomposing of active groups possessing different thermal or photochemical stability, a formation of block copolymers is possible:

$$RO{-}O{-}C(=O){-}R_1{-}N{=}N{-}R_1{-}C(=O){-}O{-}OR \xrightarrow{MMA} PMMA{\sim}O{-}C(=O){-}R_1{-}N{=}N{-}R_1{-}C(=O){-}O{\sim}PMMA \xrightarrow{Styrene} PMMA{\sim\sim}O{-}C(=O){-}R_1{-\sim\sim}PSt \qquad (2.93)$$

Such types of polyfunctional initiators mostly contain azo- or peroxy groups and may be classified as bi- and polyperoxides, bi- and polyazo derivatives, and azo peroxides. The block copolymers obtained by this method are always contaminated by homopolymer and require additional purification via extraction.

Combination of ionic and radical polymerization opens opportunities for combining various types of monomers in one molecule. For example, such procedure may be applied to specific systems with cationic polymerizable monomers (THF, cyclohexene oxide, epichlorohydrin) and the free radical polymerizable monomers (styrene, methacrylates), respectively [122, 123]:

$$Cl{-}C(=O){-}R_1{-}N{=}N{-}R_1{-}C(=O){-}Cl \xrightarrow[AgBF_4,\ -AgCl]{THF} (THF^{+}){\sim}O{-}C(=O){-}R_1{-}N{=}N{-}R_1{-}C(=O){-}O{\sim}(^{+}THF)\ \ BF_4^-\ \ BF_4^- \xrightarrow{Styrene} (THF^{+})BF_4^-{\sim\sim}O{-}C(=O){-}R_1{-\sim\sim}PSt \qquad (2.94)$$

The development of various methods of controlled polymerization significantly expanded the scope of multimode polymerization. It made possible to obtain well-defined block copolymers with desired molecular weight and composition. Combination of nitroxide-mediated and cationic polymerization allowed synthetizing block copolymers based on vinyl ethers acting as polymer surfactants and antibacterial agents [124]. The alkoxyamines containing carboxylic group are capable to initiate as living cationic polymerization (in conjunction with Lewis acid) so controlled radical polymerization (through decomposition via labile C-O). Such compounds may be successfully applied for the synthesis of various block copolymers:

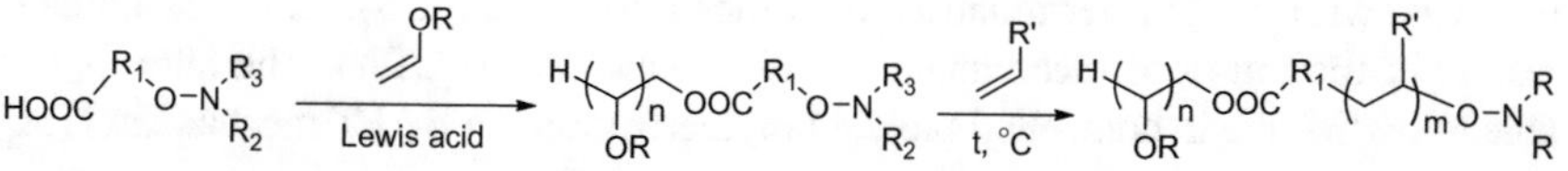

(2.95)

Living cationic polymerization may be combined with other CRP techniques, for example with ATRP [125]. The macroinitiators used in this case contain labile carbon–halogen bond capable to interact with metal halide to generate active radical species. In some cases, controlled cationic and radical processes may be conducted simultaneously in one reaction pot as in case of block copolymerization of ε-caprolactone with methyl methacrylate [tao2015].

Polyfunctional initiators may also include metal centers active in ring-opening metathesis polymerization (ROMP). Combination of metathesis and cationic ring-opening polymerizations was successfully applied for one-pot synthesis of poly(norbornene)-block-poly(lactide) copolymers [126]. The use of bifunctional poly(ethylene glycol) macroinitiator allowed to obtain ABC triblock copolymers, a perspective material for novel nanoporous membranes [127]:

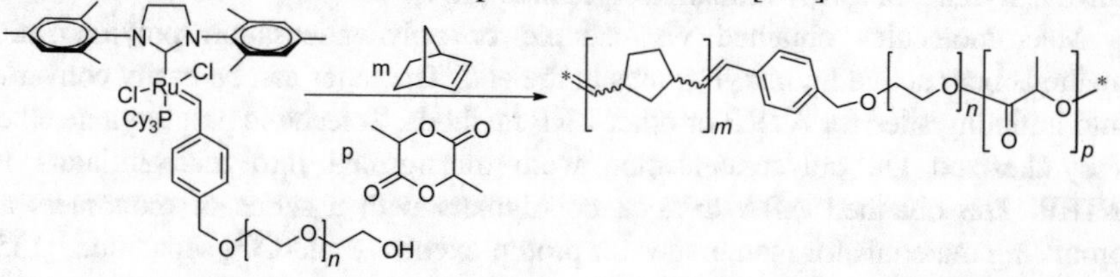

(2.96)

In turn, ROMP may be also combined with controlled radical polymerization techniques to obtain various block copolymers. A dual initiator proposed by Grubbs et al. [128] may be applied for simultaneous ROMP and ATRP, while some unsaturated alkoxyamines show high performance in dual ROMP-NMP polymerization [129].

Considerable progress in area development of controlled polymerization techniques resulted in an exponential growth of the use of dual initiators. The scope of application of dual and heterofunctional initiators is virtually unlimited due to the wide possibilities for a combination of different techniques in various ways.

(b) Multimode polymerization via end-group transformation

The described multimode initiator technique is limited by stability of each initiating center in polymerization conditions. In some cases, the synthesis of such initiators is also a challenging task. Another way of conducting multimode polymerization is a so-called end-group transformation approach. In this case, the propagating center of the first block is converted into a functionality that is capable of initiating the polymerization of the second monomer.

Such a way is very promising for a combination of various techniques with living anionic polymerization. For example, conducting of living polymerization of isoprene with further termination by diphenylethylene, CS_2, and organohalide results in formation of macrochain transfer agent (macro-CTA). The latter is capable to participate in controlled radical polymerization via RAFT mechanism [130]:

(2.97)

Using similar strategies, synthesis of various block copolymers has been also achieved, e.g., poly(ethylene-co-butylene)-block-poly(styrene-co-maleic anhydride), [131] poly(lactic acid)-block-poly(methyl methacrylate) [132], poly(lactic acid)-block-poly(*N*-isopropylacrylamide) [133]. The described synthetic approach may be applied for a combination of cationic polymerization with RAFT. So, the polythiourethane obtained by controlled cationic ring-opening polymerization was converted into macro-CTA and used in preparation of block copolymers with MMA, styrene, and *N*,*N*-dimethylacrylamide [134].

Macromolecules obtained via cationic or polycondensation polymerization methods bear active hydroxyl groups at the end. The latter can be easily converted into initiating sites for ATRP or other CRP methods. Telechelic poly(arylene ether) s synthesized via polycondensation were transformed into macronitiators for ATRP. The obtained ABA triblock copolymers with a series of monomers are promising materials for membranes for proton-exchange and CO_2 separation. [135].

Polymers obtained by ATRP contain halogen atom at the active end of macromolecules and in turn may be converted into macroinitiators for other controlled polymerization techniques, for example via "click" reactions [136]:

$$P_n{-}Br \xrightarrow{NaN_3} P_n{-}N_3 \longrightarrow P_n{-}N(\text{triazole}){-}X \qquad (2.98)$$

Using this method, poly(2-(dimethylamino)ethyl methacrylate) obtained by ATRP was successfully chain-extended by ε-caprolactone via CROP mechanism [137]. Macromolecules bearing azide end groups may be also used as a building blocks in synthesis of star copolymers via arm-first method. In this case, polymer chains obtained by controlled polymerization "clicked" to the core leading to well-defined macrostructures [138].

The described examples of multimode polymerization clearly illustrate its power for producing well-defined copolymers with different compositions and architectures. It allows combining all possible monomer units in one polymer chain leading to vast variety of novel polymer materials with unique properties demanded in high-tech industries.

2.9 Summary and Future Directions

Synthetic chemistry of polymers has passed a long path of growth and developing. Starting from laboratory scale polymerization experiments, it has grown to routine industrial mean of polymer production from one side and to powerful tool for direct synthesis of novel well-defined materials from another. The dominant role in development of new effective ways to polymers with desired properties belongs to catalytic systems mostly based on transition metals and methods of "living" polymerization (living cationic, anionic, radical, and ring-opening polymerization).

Living polymerization conquers a firm place in the modern polymer market. Starting from pioneer manufactures by DuPont and Daikin, it was picked up and developed by other leaders of chemical industry such as Arkema, Ciba, Kaneka, EKFA, BYK, Rhodia. Today, living methods of polymerization are used for industrial production of pigments, sealants, emulsion stabilizers, surfactants, block copolymers with desired properties including photoresists.

At the same time, the number of industrial products obtained by controlled polymerization methods remains small in comparison with overall bulk of polymer production. It is determined by a scope of different reasons including high reagent costs and purification necessity, lower polymerization rate, and rather high temperatures for effective polymerization. The founding of happy medium between product quality and its cost is a key factor for successful production of novel product into industrial scale [50]. The synergetic use of experimental data on polymerization mechanisms, including strict kinetic studies, computer modeling, and native catalyst testing, will result in development of novel effective catalytic systems for controlled polymerization. In this case, the search of novel more effective and convenient systems is a challenging task in the area of controlled polymerization.

References and Further Readings

1. Schluter DA, Hawker C, Sakamoto J (eds) (2012) Synthesis of polymers: new structures and methods. Wiley-VCH Verlag & Co., Weinheim
2. Moad G, Solomon DH (2006) The chemistry of radical polymerization, 2nd edn. Elsevier Science Inc., New York
3. Brandrup J, Immergut EH, Grulke EA (eds) (1999) Polymer handbook, 4th edn. Wiley, New York
4. Grishin DF (1993) Coordination-radical (co)polymerisation of vinyl monomers in the presence of organic compounds of Group III-V elements. Russ Chem Rev 10:951–962. https://doi.org/10.1070/RC1993v062n10ABEH000056
5. Matyjaszewski K (1998) Overview: fundamentals of controlled/living radical polymerization. In: Matyjaszewski K (ed) Controlled radical polymerization. ACS Symposium Series. vol 685. American Chemical Society, Washington, DC, p 2
6. Odian G (2004) Principles of polymerization, 4th edn. Wiley, Hoboken
7. Trommsdorff E, Kohle H, Lagally P (1948) Zur polymerisation des methacrylsäuremethylesters. Makromol Chem 1(3):169–198. https://doi.org/10.1002/macp.1948.020010301

8. Norrish RGW, Smith RR (1942) Catalysed polymerization of methyl methacrylate in the liquid phase. Nature 150:336–337
9. Russell GT (1994) On exact and approximate methods of calculating an overall termination rate coefficient from chain length dependent termination rate coefficients. Macromol Theor Simul 3:439–468. https://doi.org/10.1002/mats.1994.040030213
10. Wang X, Ruckenstein EJ (1993) CO_2 reforming of CH_4 over Co/MgO solid solution catalysts—effect of calcination temperature and Co loading. Appl Polym Sci 49:2179–2188
11. Grishin DF, Grishin ID (2015) Radical-initiated controlled synthesis of homo- and copolymers based on acrylonitrile. Russ Chem Rev 84:712–736. https://doi.org/10.1070/RCR4476
12. Braun D, Cherdron H, Rehahn M, Ritter H, Voit B (2013) Polymer synthesis: theory and practice fundamentals, methods, experiments, 5th edn. Springer, Heidelberg
13. Szwarc M (1956) «Living» polymers. Nature 176:1168
14. Matyjaszewski K, Spanswick J (2005) Controlled/living radical polymerization. Mater Today 8(3):26–33. https://doi.org/10.1016/S1369-7021(05),00745-5
15. Matyjaszewski K, Tsarevsky NV (2014) Macromolecular engineering by atom transfer radical polymerization. J Am Chem Soc 136:6513–6533. https://doi.org/10.1021/ja408069v
16. Destarac M (2010) Controlled radical polymerization: industrial stakes, obstacles and achievements. Macromol React Eng 4:165–179. https://doi.org/10.1002/mren.200900087
17. Grishin ID, Grishin DF (2011) Controlled radical polymerization: prospect for application and industrial synthesis of polymers. Russ J Appl Chem 84(12):2021–2033
18. Matyjaszewski K (2012) Atom transfer radical polymerization (ATRP): current status and future perspectives. Macromolecules 45:4015–4428. https://doi.org/10.1021/ma3001719
19. Moad G, Thang ESH (2013) Chem Asian J 8:1634
20. Nicolas J, Guillaneuf Y, Lefay C, Bertin D, Gigmes D, Charleux B (2013) Nitroxide-mediated polymerization. Prog Polymer Sci 38:63–235. https://doi.org/10.1016/j.progpolymsci.2012.06.002
21. Hawker CJ, Bosman AW, Harth E (2001) New Polymer synthesis by nitroxide mediated living radical polymerizations. Chem Rev 101:3661–3688. https://doi.org/10.1021/cr990119u
22. Kolyakina EV, Grishin DF (2009) Nitroxide radicals formed in situas polymer chain growth regulators. Russ Chem Rev 78:535–568. https://doi.org/10.1070/RC2009v078n06ABEH004026
23. Poli R (2006) Relationship between one-electron transition-metal reactivity and radical polymerization processes. Angew Chem Int Ed 45:5058–5070. https://doi.org/10.1002/anie.200503785
24. Allan LEN, Perry MR, Shaver MP (2012) Organometallic mediated radical polymerization. Prog Polymer Sci 37:127–156. https://doi.org/10.1016/j.progpolymsci.2011.07.004
25. Chong YK, Le TPT, Moad G, Rizzardo E, Thang SH (1999) A more versatile route to block copolymers and other polymers of complex architecture by living radical polymerization: the RAFT process. Macromolecules 32:2071–2074. https://doi.org/10.1021/ma981472p
26. Debuigne A, Poli R, Jérôme C, Jérôme R, Detrembleur C (2009) Overview of cobalt-mediated radical polymerization: roots, state of the art and future prospects. Prog Polymer Sci 34:211–239. https://doi.org/10.1016/j.progpolymsci.2008.11.003
27. Maria S, Kaneyoshi H, Matyjaszewski K, Poli R (2007) Effect of electron donors on the radical polymerization of vinyl acetate mediated by [Co(acac)2]: degenerative transfer versus reversible homolytic cleavage of an organocobalt(III) complex. Chem Eur J 13:2480–2492. https://doi.org/10.1002/chem.200601457
28. Koumura K, Satoh K, Kamigaito M, Okamoto Y (2006) iodine transfer radical polymerization of vinyl acetate in fluoroalcohols for simultaneous control of molecular weight, stereospecificity, and regiospecificity. Macromolecules 39:4054–4061. https://doi.org/10.1021/ma0602775
29. Poli R (2015) New phenomena in organometallic-mediated radical polymerization (OMRP) and perspectives for control of less active monomers. Chem Eur J 21:6988–7001. https://doi.org/10.1002/chem.201500015

30. Kermagoret A, Jérôme C, Detrembleur C, Debuigne A (2015) In situ bidentate to tetradentate ligand exchange reaction in cobalt-mediated radical polymerization. Eur Polymer J 62:312–321. https://doi.org/10.1016/j.eurpolymj.2014.08.003
31. Stoffelbach F, Poli R, Maria S, Richard P (2007) How the interplay of different control mechanisms affects the initiator efficiency factor in controlled radical polymerization: an investigation using organometallic MoIII-based catalysts. J Organometal Chem 692:3133–3143. https://doi.org/10.1016/j.jorganchem.2006.11.031
32. Stoffelbach F, Poli R, Richard P (2003) Half-sandwich molybdenum(III) compounds containing diazadiene ligands and their use in the controlled radical polymerization of styrene. J Organometal Chem 663:269–276. https://doi.org/10.1016/S0022-328X(02)01878-8
33. Shchepalov AA, Grishin DF (2008) Dicyclopentadienyltitanium chlorides as regulators of free-radical polymerization of vinyl monomers. Polymer Sci A Polymer Chem 50(4): 382–387. https://doi.org/10.1134/S0965545X08040044
34. Schroeder H, Lake BRM, Demeshko S, Shaver MP, Buback M (2015) A synthetic and multispectroscopic speciation analysis of controlled radical polymerization mediated by amine–bis(phenolate)iron complexes. Macromolecules 48:4329–4338. https://doi.org/10.1021/acs.macromol.5b01175
35. Hawker CJ, Bosman AW, Harth E (2001) New polymer synthesis by nitroxide mediated living radical polymerizations. Chem Rev 101:3661–3688. https://doi.org/10.1021/cr990119u
36. Vinas J, Chagneux N, Gigmes D, Trimaille T, Favier A, Bertin D (2008) SG1-based alkoxyamine bearing a N-succinimidyl ester: a versatile tool for advanced polymer synthesis. Polymer 49:3639–3647. https://doi.org/10.1016/j.polymer.2008.06.017
37. Moad G, Rizzardo E, Thang SH (2013) RAFT polymerization and some of its applications. Chem Asian J 8:1634–1644. https://doi.org/10.1002/asia.201300262
38. Wang J-S, Matyjaszewski K (1995) Controlled/"living" radical polymerization. Atom transfer radical polymerization in the presence of transition-metal complexes. J Am Chem Soc 117:5614–5615. https://doi.org/10.1021/ja00125a035
39. Kato M, Kamigaito M, Sawamoto M, Higashimura T (1995) Polymerization of methyl methacrylate with the carbon tetrachloride/dichlorotris- (triphenylphosphine)ruthenium(ii)/ methylaluminum bis(2,6-di-tert-butylphenoxide) initiating system: possibility of living radical polymerization. Macromolecules 28:1721–1723. https://doi.org/10.1021/ma00109a056
40. Fujimura K, Ouchi M, Sawamoto M (2015) Ferrocene cocatalysis for iron-catalyzed living radical polymerization: active, robust, and sustainable system under concerted catalysis by two iron complexes. Macromolecules 48:4294–4300. https://doi.org/10.1021/acs.macromol.5b00836
41. Poli R, Allan LEN, Shaver MP (2014) Iron-mediated reversible deactivation controlled radical polymerization. Prog Polymer Sci 39:1827–1845. https://doi.org/10.1016/j.progpolymsci.2014.06.003
42. De Roma A, Yanga H-J, Milione S, Capacchione C, Roviello G, Grassi A (2011) Atom transfer radical polymerization of methylmethacrylate mediated by a naphtyl–nickel(II) phosphane complex. Inorg Chem Commun 14:542–544. https://doi.org/10.1016/j.inoche.2011.01.019
43. Trotta JT, Fors BP (2016) Organic catalysts for photocontrolled polymerizations. Synlett 27:702–713. https://doi.org/10.1055/s-0035-1561264
44. Treat NJ, Sprafke H, Kramer JW, Clark PG, Barton BE, de Alaniz JR, Fors BP, Hawker CJ (2014) Metal-free atom transfer radical polymerization. J Am Chem Soc 136:16096–16101. https://doi.org/10.1021/ja510389m
45. Pan X, Lamson M, Yan J, Matyjaszewski K (2015) Photoinduced metal-free atom transfer radical polymerization of acrylonitrile. ACS Macro Lett 4:192–196. https://doi.org/10.1021/mz500834g
46. Treat NJ, Fors BP, Kramer JW, Christianson M, Chiu CY, de Alaniz JR, Hawker CJ (2014) Controlled radical polymerization of acrylates regulated by visible light. ACS Macro Lett 3:580–584. https://doi.org/10.1021/mz500242a

47. Miyake GM, Theriot JC (2014) Perylene as an organic photocatalyst for the radical polymerization of functionalized vinyl monomers through oxidative quenching with alkyl bromides and visible light. Macromolecules 47:8255–8261. https://doi.org/10.1021/ma502044f
48. Xu J, Shanmugam S, Duong HT, Boyer C (2015) Organo-photocatalysts for photoinduced electron transfer-reversible addition–fragmentation chain transfer (PET-RAFT) polymerization. Polymer Chem 6:5615–5624. https://doi.org/10.1039/C4PY01317D
49. Shanmugam S, Boyer C (2016) Metal-free catalysts enable synthesis of polymers for biomedical and electronics applications. Science 352:1053–1054. https://doi.org/10.1126/science.aaf7465
50. Matheson RR (2000) The commercialization of controlled polymer synthesis. The Knowledge Foundation, Cambridge
51. Szwarc M (1968) Carbanions, living polymers and electron transfer processes. Interscience, NewYork
52. Bywater S (1985) Anionic Polymerization. In: Mark HF, Bikales NM, Overberger CG, Menges G (eds) Encyclopedia polymer science & engineering, vol 2. Wiley-Interscience, NewYork, p 1
53. Szwarc M, Levy M, Milkovich R (1956) Polymerization initiated by electron transfer to monomer. A new method of formation of block polymers. J Am Chem Soc 78:2656–2657. https://doi.org/10.1021/ja01592a101
54. Hadjichristidis N, Pitsikalis M, Pispas S, Iatrou H (2001) Polymers with complex architecture by living anionic polymerization. Chem Rev 101:3747–3792. https://doi.org/10.1021/cr9901337
55. Scott ND (1939) Method of polymerization. US Patent 2,181,771, 28 Nov 1939
56. Welch FJ (1960) Polymerization of Styrene by n-Butyllithium. II. Effect of Lewis acids and bases. J Am Chem Soc 82:6000–6005. https://doi.org/10.1021/ja01508a009
57. Burnett GM, Young RN (1966) The polymerization of substituted styrenes by butyl lithium—I. The initiation reaction. Eur Polym J 2:329–338. https://doi.org/10.1016/0014-3057(66)90014-0
58. Eberhardt GC, Butte WA (1964) A catalytic telomerization reaction of ethylene with aromatic hydrocarbons. J Org Chem 29:2928–2932. https://doi.org/10.1021/jo01033a029
59. Langer AW (1965) Reactions of chelated organolithium compounds. Trans NY Acad Sci 27:741–747. https://doi.org/10.1111/j.2164-0947.1965.tb02233.x
60. Tobolsky AV, Rogers CE (1959) Isoprene polymerization by organometallic compounds. II. J Polym Sci 40:73–89. https://doi.org/10.1002/pol.1959.1204013605
61. Gourdenne A, Sigwalt P (1967) Stability of the living polymers of dienes in relation with the preparation of block copolymers. Eur Polym J 3(3):481–499. https://doi.org/10.1016/0014-3057(67)90016-X
62. Guyot A, Vialle J (1970) Isoprene polymerization by butyllithium in cyclohexane. II. propagation reaction. J Macromol Sci A4:107–125. https://doi.org/10.1080/00222337008060968
63. Tobolsky AV, Boudreau RJ (1961) Ionic copolymerization of substituted styrenes. J Polym Sci 51:S53–S56. https://doi.org/10.1002/pol.1961.1205115632
64. Worsfold DJ, Bywater S (1964) Anionic polymerization of isoprene. Can J Chem 42:2884–2892. https://doi.org/10.1139/v64-426
65. Hsieh H, Kelley DJ, Tobolsky AV (1957) Polymerization of isoprene with lithium dispersions and lithium alkyls using tetrahydrofuran as solvent. J Polym Sci 26:240–242. https://doi.org/10.1002/pol.1957.1202611315
66. Lohr G (1974) Schulz GV (1974) Kinetics of anionic polymerization of methylmetacrylate with caesium and sodium as counterions in tetrahydrofuran. Eur Polym J 10:121–130. https://doi.org/10.1016/0014-3057(74)90077-9
67. Muller AHE, Hocker H, Schulz GV (1977) Rate constants of the tactic monomer addition in the anionic polymerization of methyl methacrylate in THF with cesium as counterion. Macromolecules 10:1086–1089. https://doi.org/10.1021/ma60059a037

68. Freyss D, Rempp P, Benoit H (1964) Polydispersity of anionically prepared block copolymers. J Polym Sci Polym Lett 2:217–222. https://doi.org/10.1002/pol.1964.110020214
69. Bailey JT, Bishop ET, Hendricks WR, Holden G, Legge NR (1966) Thermoplastic elastomers. Physical properties and applications. Rubber Age 98(10):69–74
70. Hadjichristidis N, Pitsikalis M, Pispas S, Iatrou H (2001) Polymers with complex architecture by living anionic polymerization. Chem Rev 101:3747–3792. https://doi.org/10.1021/cr9901337
71. Mayr H, Kempf B, Ofial AR (2003) π-nucleophilicity in carbon–carbon bond-forming reactions. Acc Chem Res 36(1):66–77. https://doi.org/10.1021/ar020094c
72. Mayr H (1999) Rate constants and reactivity ratios in carbocationic polymerizations. Ionic polymerization and related processing. NATO science series E, vol 359. Kluwer, Dordecht, p 99
73. Kennedy JP (1975) Cationic polymerization of olefins: a critical inventory. Wiley-Interscience, New York
74. Kennedy JP, Marechal E (1982) Carbocationic polymerization. Wiley-Interscience, New York, p 85
75. Chang VSC, Kennedy JP, Ivan B (1980) New telechelic polymers and sequential copolymers by polyfunctional initiator-transfer agents (inifers). Polym Bull 3(6):339–346. https://doi.org/10.1007/BF00255093
76. Russell R, Moreau M, Charleux B, Vairon JP, Matyjaszewski K (1998) Stopped-Flow and 1H NMR study of the ionization of cumyl chloride by boron trichloride. Macromolecules 31:3775–3782. https://doi.org/10.1021/ma971883q
77. Kennedy JP, Ivan B (1992) Designed polymers by carbocationic macromolecular engineering. Theory and practice. Hanser Publishers, Munich, p 173
78. Ziegler K, Holzkamp E, Breil H, Martin H (1955) Polymerisation von Äthylen und anderen Olefinen. Angew Chem 67(16):426–426
79. Ziegler K (1965) Consequences and development of an invention. rubber chemistry and technology. Rubber Chem Technol 38:23–36. https://doi.org/10.5254/1.3535634
80. Natta G (1956) Stereospezifische katalysen und isotaktische polymere. Angew Chem 68:393–403
81. Natta G, Danusso F, Sianesi D (1959) Structure and reactivity of vinyl aromatic monomers in coordinated anionic polymerization and copolymerization. Makromol Chem 30:238–246. https://doi.org/10.1002/macp.1959.020300116
82. Beerman C, Bestian H (1959) Metallorganische Titan-Verbindungen als Polymerisationskatalysatoren. Angew Chem 71:618–623. https://doi.org/10.1002/ange.19590711908
83. Yermakov N, Zakharov V (1975) One component catalyst for olefin polymerization. Adv Catal 24:173–219
84. Pino P, Mulhauft R (1980) Stereospecific polymerization of propylene: an outlook 25 years after its discovery. Angew Chem Int Ed 19:857–875. https://doi.org/10.1002/anie.19800857
85. Zambelli A, Locatelli P, Rigamonti E (1979) Carbon-13 nuclear magnetic resonance analysis of tail-to-tail monomeric units and of saturated end groups in polypropylene. Macromolecules 12:156–159. https://doi.org/10.1021/ma60067a034
86. Zambelli A, Allegra G (1980) Reaction mechanism for syndiotactic specific polymerization of propene. Macromolecules 13:42–49. https://doi.org/10.1021/ma60073a008
87. Asakura T, Ando I, Nishioka A, Doi Y and Keii T (1977) ^{13}C NMR analysis of chemical inversion in polypropylene. Makromol Chem 178:791–801. https://doi.org/10.1002/macp.1977.021780316
88. Doi Y, Asakura T (1975) Catalytic regulation for isotactic orientation in propylene polymerization with Ziegler-Natta catalyst. Makromol Chem 176:507–509. https://doi.org/10.1002/macp.1975.021760221

89. Zambelli A, Giongo M, Natta G (1968) Polymerization of propylene to syndiotactic polymer. IV. Addition to the double bond. Makromol Chem 112:183–196. https://doi.org/10.1002/macp.1968.021120116
90. Natta G (1958) Stereospecific polymerizations by means of coordinated anionic catalysis: introductory lecture. J Inorg Nucl Chem 8:589–611. https://doi.org/10.1016/0022-1902(58)80234-1
91. Pino P, Oschwald A, Ciardelli F, Carlini C, Chiellini E (1975) Stereoselection and stereoelection in α-olefin polymerization. In: Chien JCW (ed) Coordination polymerization. Academic Press, New York, pp 25–72
92. Arlman EG, Cossee P (1964) Ziegler-Natta catalysis III. Stereospecific polymerization of propene with the catalyst system $TiCl_3$-$AlEt_3$. J Catal 3:99–104. https://doi.org/10.1016/0021-9517(64)90097-1
93. Natta G, Pasquon I (1959) The kinetics of stereospecific polymerization of olefins. Adv Catal 11:1–68
94. Boor J (1979) Ziegler-Natta catalysts and polymerizations. Academic Press, New York
95. Sinn H, Kaminsky W, Vollmer HJ (1980) "Living polymers" on polymerization with extremely productive Ziegler catalysts. Angew Chem Int Ed 19:390–392. https://doi.org/10.1002/anie.198003901
96. Kaminsky W, Sinn H (2013) Methylaluminoxane: key component for new polymerization catalysts. Adv Polym Sci 258:1–28. https://doi.org/10.1007/12_2013-226
97. Kaminsky W (2004) The discovery of metallocene catalysts and their present state of the art. J Polym Sci Polym Chem Ed 42:3911–3921. https://doi.org/10.1002/pola.20292
98. Sinn H (1995) Proposals for structure and effect of methylalumoxane based on mass balances and phase separation experiments. Macromol Symp 97:27–52. https://doi.org/10.1002/masy.19950970105
99. Kaminsky W, Spiehl R (1989) Copolymerization of cycloalkenes with ethylene in presence of chiral zirconocene catalysts. Makromol Chem 190:515–526. https://doi.org/10.1002/macp.1989.021900308
100. Ewen JA, Jones RL, Razavi A, Ferrara JP (1988) Syndiospecific propylene polymerizations with Group IVB metallocenes. J Am Chem Soc 110:6255–6256. https://doi.org/10.1021/ja00226a056
101. Abu-Surrah AS, Rieger B (1996) Late transition metal complexes: catalysts for a new generation of organic polymers. Angew Chem Int Ed 35:2475–2477. https://doi.org/10.1002/anie.199624751
102. Ittel SD, Johnson LK, Brookhart M (2000) Late-metal catalysts for ethylene homo- and copolymerization. Chem Rev 100:1169–1204. https://doi.org/10.1021/cr9804644
103. Small BL, Brookhart M, Bennett AMA (1998) Highly active iron and cobalt catalysts for the polymerization of ethylene. J Am Chem Soc 120:4049–4050. https://doi.org/10.1021/ja9802100
104. Britovsek GJP, Gibson VC, Wass DF (1999) The search for new-generation olefin polymerization catalysts: life beyond metallocenes. Angew Chem Int Ed 38:428–447. https://doi.org/10.1002/(SICI)1521-3773(19990215)38:4%3c428:AID-ANIE428%3e3.0.CO;2-3
105. Gibson VC, Wass DF (1999) Olefin polymerization catalysts. Chem Br 7:20–23
106. Zambelli A, Dipietro J, Gatti G (1963) The nature of active components in catalytic systems prepared from $TiCl_3$, monoalkylaluminum dihalides, and electron-donor substances, in the polymerization of propylene. J Polym Sci 1:403–409. https://doi.org/10.1002/pol.1963.100010136
107. Doi Y, Takada M, Keii T (1979) Molecular weight distribution and kinetics of low-temperature propene polymerization with soluble vanadium-based ziegler catalysts. Bull Chem Soc Jpn 52:1802–1806. https://doi.org/10.1246/bcsj.52.1802

108. Doi Y, Ueki S, Keii T (1979) "Living" coordination polymerization of propene initiated by the soluble V(acac)$_3$-Al(C_2H_5)$_2$Cl system. Macromolecules 12:814–819. https://doi.org/10.1021/ma60071a004
109. Fukui Y, Murata M, Soga K (1999) Living polymerization of propylene and 1-hexene using bis-Cp type metallocene catalysts. Macromol Rapid Commun 20:637–640. https://doi.org/10.1002/(SICI)1521-3927(19991201)20:12%3c637:AID-MARC637%3e3.0.CO;2-N
110. Slomkowski S, Duda A (1993) Anionic ring-opening polymerization. In: Brunelle DJ (ed) Ringopening polymerization. Mechanisms, catalysis, structure, utility. Hanser, Munich, p 87
111. Natta G, Dall'Asta G (1969) Elastomers from cyclic olefins. In: Kennedy JP, Tornquist EGM (eds) Polymer chemistry of synthetic elastomers, part II, High polymer series, vol 23. Interscience, New York, p. 703
112. Grubbs RH (2003) Handbook of metathesis, vol 3. Wiley-VCH, Weinheim
113. Calderon N (1972) Olefin metathesis reaction. Acc Chem Res 5:127–132. https://doi.org/10.1021/ar50052a002
114. Calderon N, Ofstead EA, Ward JP, Judy WA, Kenneth W (1968) Scott Olefin metathesis. I. Acyclic vinylenic hydrocarbons. J Am Chem Soc 90:4133–4140. https://doi.org/10.1021/ja01017a039
115. Wallace KC, Liu AH, Dewan JC, Schrock AR (1988) Preparation and reactions of tantalum alkylidene complexes containing bulky phenoxide or thiolate ligands. Controlling ring-opening metathesis polymerization activity and mechanism through choice of anionic ligand. J Am Chem Soc 110:4964–4977. https://doi.org/10.1021/ja00223a014
116. Khosravi E, Feast WJ, Al-Hajaji AA, Leejarkpai T (2000) ROMP of *n*-alkyl norbornene dicarboxyimides: from classical to well-defined initiators, an overview. J Mol Cat A: Chem 160:1–11. https://doi.org/10.1016/S1381-1169(00),00227-2
117. Bielawski CW, Grubbs RH (2007) Living ring-opening metathesis polymerization Prog Polym Sci 32:1–29. https://doi.org/10.1016/j.progpolymsci.2006.08.006
118. Ivin KJ, Mol JC (1997) Olefin metathesis and metathesis polymerization. Academic Press, San Diego
119. Andruzzi F, Pilcher G, Virmani Y, Plesch PH (1977) Enthalpy of polymerisation of 7-oxabicyclo[2.2.1]heptane, and *exo-* and *endo*-2-methyl-7-oxabicyclo[2.2.1]heptanes. Makromol Chem 178:236–2373. https://doi.org/10.1002/macp.1977.021780822
120. Bernaerts KV, Du Prez FE (2006) Dual/heterofunctional initiators for the combination of mechanistically distinct polymerization techniques. Prog Polym Sci 31:671–722. https://doi.org/10.1016/j.progpolymsci.2006.08.007
121. CrI S, Comăniţă E, Păstrăvanu M, Dumitriu S (1986) Progress in the field of bi- and poly-functional free-radical polymerization initiators. Prog Polym Sci 12:1–109. https://doi.org/10.1016/0079-6700(86)90006-7
122. Yagoi Y, Hizal G, Önen A, Serhatli I (1994) Synthetic routes to block copolymerization by changing mechanism from cationic polymerization to free radical polymerization. Macromol Symp 84:127–136. https://doi.org/10.1002/masy.19940840116
123. Mishra KM (1996) Synthesis of polyisobutylene-based macroinitiators and block copolymers via multimode polymerization. Macromolecules 29:5228–5230. https://doi.org/10.1021/ma9603185
124. Le D, Phan TNT, Autissier L, Charles L, Gigmes D (2016) Well-defined block copolymer synthesis via living cationic polymerization and nitroxide-mediated polymerization using carboxylic acid-based alkoxyamines as dual initiator. Polym Chem 8:1659–1667. https://doi.org/10.1039/C5PY01934F
125. Bernaerts KV, Du Prez FE (2005) Design of novel poly(methyl vinyl ether) containing AB and ABC block copolymers by the dual initiator strategy. Polymer 46:8469–8482. https://doi.org/10.1016/j.polymer.2005.01.103

126. Jung H, Brummelhuis N, Yang SK, Weck M (2013) One-pot synthesis of poly(norbornene)-block-poly(lactide) copolymers using a bifunctional initiator. Poly Chem 4:2837–2840. https://doi.org/10.1039/c3py21067g
127. Freudensprung I, Klapper M, Müllen K (2015) Triblock Terpolymers by simultaneous tandem block polymerization (STBP) Macromol. Rapid Comm 37:209–214. https://doi.org/10.1002/marc.201500568
128. Bielawski CW, Louie J, Grubbs RH (2000) Tandem catalysis: three mechanistically distinct reactions from a single ruthenium complex. J Am Chem Soc 122:12872–12873. https://doi.org/10.1021/ja001698j
129. Miura Y, Sakai Y, Taniguchi I (2003) Syntheses of well-defined poly(siloxane)-b-poly (styrene) and poly(norbornene)-b-poly(styrene) block copolymers using functional alkoxyamines. Polymer 44:603–611. https://doi.org/10.1016/S0032-3861(02)00785-1
130. Zhang C, Yang Y, He J (2013) Direct transformation of living anionic polymerization into RAFT based polymerization. Macromolecules 46:3985–3994. https://doi.org/10.1021/ma4006457
131. Brouwer HD, Schellekens MAJ, Klumperman B, Monteiro M, German AL (2000) Controlled radical copolymerization of styrene and maleic anhydride and the synthesis of novel polyolefin-based block copolymers by reversible addition–fragmentation chain-transfer (RAFT) polymerization. J Polym Sci Polym Chem 38:3596–3603. https://doi.org/10.1002/1099-0518(20001001)38:19%3c3596:AID-POLA150%3e3.0.CO;2-F
132. Perrier S, Takolpuckdee P, Westwood J, Lewis DM (2004) Versatile chain transfer agents for reversible addition fragmentation chain transfer (RAFT) polymerization to synthesize functional polymeric architectures. Macromolecules 37:2709–2717. https://doi.org/10.1021/ma035468b
133. Hales M, Barner-Kowollik T, Davis TP, Stenzel MH (2004) Shell-cross-linked vesicles synthesized from block copolymers of poly(D, L-lactide) and poly(*N*-isopropyl acrylamide) as thermoresponsive nanocontainers. Langmuir 20:10809–10817. https://doi.org/10.1021/la0484016
134. Nagai A, Hamaguchi T, Kikukawa K, Kawamoto E, Endo T (2007) Synthesis of polythiourethane-based macro chain transfer agents and their block copolymers with vinyl monomers via controlled multimode polymerization. Macromolecules 40:6454–6456. https://doi.org/10.1021/ma071079w
135. Agudelo NA, Elsen AM, He H, Lopez BL, Matyjaszewski K (2014) ABA Triblock copolymers from two mechanistic techniques: polycondensation and atom transfer radical polymerization. J Polym Sci Polym Chem 53:228–238. https://doi.org/10.1002/pola.27300
136. Coessens V, Matyjaszewski K (1999) End group transformation of polymers prepared by ATRP, substitution to azides. J Macromol Sci, Pure Appl Chem 36:667–679. https://doi.org/10.1081/MA-100101556
137. Bruce K, Javakhishvili I, Fogelstrom L, Carlmark A, Hvilstedb S, Malmstrom E (2014) Well-defined ABA- and BAB-type block copolymers of PDMAEMA and PCL. RSC Adv. 4:25809–25818. https://doi.org/10.1039/c4ra04325a
138. Fournier D, Hoogenboom R, Schubert US (2007) Clicking polymers: a straightforward approach to novel macromolecular architectures. Chem Soc Rev 36:1369–1380. https://doi.org/10.1039/b700809k

Chapter 3
Polymer's Characterization and Properties

Olumide Bolarinwa Ayodele and Peter Adeniyi Alaba

Abstract This chapter discussed the characterization and properties of polymeric materials. Prominent characterization techniques used for analyzing polymeric materials are mass spectrometry (MS), nuclear magnetic resonance (NMR), and gel permeation chromatography (GPC), which are used for measuring mass-to-charge ratio (m/z) of analyte ions. Other methods used for this purpose include matrix-assisted laser desorption/ionization (MALDI), electrospray ionization (ESI), and secondary-ion mass spectrometry (SIMS). X-ray diffraction (XRD) is used for solid-state analysis such as degree of crystallinity and crystal structure as well as the unit cell parameters, while Fourier transform infrared spectroscopy (FTIR) is used for identification of the polymer functional groups. NMR helps to identify and characterize various polymers and also provides information on the mobility of their molecules, while X-ray photoelectron spectroscopy (XPS) provides information regarding the chemical composition of polymeric materials. The physical properties such as hydrophobicity, functional groups, and flexibility of the polymer chain structure and chemical properties such as chemical reactivity, toxicity, biocompatibility, chirality, adsorption capacities, chelation, and polyfunctionality of polymeric materials are also discussed.

Keywords Polymers · Characterization methods · Physical properties
Chemical properties

O. B. Ayodele (✉)
International Iberian Nanotechnology Laboratory,
Micro- and Nanofabrication Department, Av. Mestre José Veiga,
4715-330 Braga, Portugal
e-mail: ayodele_olumide@yahoo.com; olumide.ayodele@inl.int

P. A. Alaba
Department of Chemical Engineering, Covenant University,
P.M.B 1023, Sango-ota, Nigeria

R. Das (ed.), *Polymeric Materials for Clean Water*,
Springer Series on Polymer and Composite Materials,
https://doi.org/10.1007/978-3-030-00743-0_3

3.1 Introduction

The focus of this chapter is characterization and properties of polymeric materials. Characterization and properties of polymeric materials is a vital aspect in the analysis of organic materials. The analytic methods used for polymeric materials is based on viscoelastic properties, precisely, dynamic mechanical testing. In addition, techniques for determination of colloidal scale structure like molecular weight and chain structure for high molecular weight polymeric materials such as small angle scattering (SAS), gel permeation chromatography (GPC), and other various methods were discussed.

Polymeric materials are extremely hydrophilic; they swell intensely in aqueous solution and are insoluble in organic solvents. Due to their gel structure, polymeric materials are characterized by their mechanical and chemical stability, density, thermal transition, water retention infrared (IR) spectroscopy, and high-resolution solid-state nuclear magnetic resonance (NMR). Furthermore, since most polymeric materials are utilized in the solid state, traditional characterization techniques aimed such as thermal analysis, optical and electron microscopy, and X-ray diffraction (XRD) are usually employed. These techniques are rich channels of information regarding polymeric materials, especially for their molecular and structural mobilities. Hartmann–Hahn cross-polymerization (CP), combined magnetic angle spinning (MAS), and polar decoupling (PD) have been extensively used in the study of NMR for solids, including gels and cross-linked polymers. The technique is sensitive to different molecular mobilities and used for characterization and identification of the structure of various polymers. The individual relaxation times provide information regarding local mobility. This chapter discussed the state-of-the-art analytical techniques for polymeric materials. These include XRD, FTIR, NMR, XPS, VSM, as well as the chemical properties of biopolymer-based and magnetic polymer-based materials, which individually probe a particular aspect of the materials.

3.2 Polymer Characterizations

3.2.1 *Mass Spectrometry*

Mass spectrometry (MS) is now acknowledged as an essential polymer characterization method along with NMR and GPC. Mass spectrometer measures the mass-to-charge ratio (m/z) of analyte ions. The sample to be analyzed by a mass spectrometer must be in the gas phase and charged. MS of polymers and polymeric surfaces is a comprehensive field of ongoing polymer-based researches. Several MS techniques are used for probing polymers and polymeric surfaces including a number of polymer compositions and surfaces that needed to be investigated carefully [13]. The most popular types of MS instruments used for polymers and surfaces analysis are Fourier transform (FT, also called ion cyclotron resonance—ICR), quadrupole, time-of-flight (TOF), and magnetic sector. Recently, there is a thrilling innovation in mass spectrometer called orthogonal TOF instruments, which is used in the study of electrospray ionization (ESI) [17].

Normally, MS of polymers or surfaces seems rather unsuitable since MS techniques involve gas-phase ions for the analysis to be successful. However, polymers are made up of bulky, entwined macromolecules, which could not be readily transformed to gas-phase species. Despite this natural incompatibility of polymers and gas-phase ions, MS techniques have been designed to provide ample information about polymeric adsorbents. The usual MS techniques, such as GCMS and pyrolysis, give information regarding the contaminants, repeat units, and additives in polymeric adsorbents.

Since soft ionization techniques, such as fast atom bombardment (FAB) and field desorption (FD), were developed, the capability of MS to analyze intact oligomers of light molecular weight (LMW) polymers has been extended. However, the recent developments of matrix-assisted laser desorption/ionization (MALDI), ESI, and secondary-ion mass spectrometry (SIMS) have declined the utilization of FAB and FD. But the utilization of MS for polymer characterization has significantly increased. For instance, Li et al. [22] utilized MALDI-TOF-MS to examine the LMW chitosan and chito-oligomers prepared by enzymatic hydrolysis of chitosan. The MALDI-TOF-MS result shows that signal strengths of the chito-oligomers are similar, regardless of their structure. The MALDI-TOF-MS spectrum of LMW chitosan with molecular weight of 1.5×10^{-3} reveals that the products were mostly of chito-oligomers, particularly with degree of polymerization (DP) ranging from 3 to 8 [23]. Along with the techniques developed to analyze bulk polymers, SIMS has been developed and applied to characterize polymer surfaces.

3.2.2 Gel Permeation Chromatography (GPC)

GPC also referred to as size-exclusion chromatography (SEC), which is used in the determination of number average MW (M_n), weight-average molecular weight (M_w), and MW dispersion (M_w/M_n) of polymers [22]. The limitations of the GPC techniques include the fact that the elution volumes have to be calibrated to the MW for different polymers, and in addition, the process exhibits a very low mass resolution. On the other hand, MALDI technique exhibits a high mass resolution, high sensitivity, and high mass accuracy, but it discriminates samples and exhibits incorrect average MW for samples with extensive polydispersity [13]. Coupling GPC with MALDI is a normal synergistic approach toward maximizing the performance of these techniques for polymers characterization. This combination could be achieved in three different ways:

(a) Individual collection of the GPC elution fractions and subsequent usage of MALDI technique to analyze the fractions, thereby improving calibration for GPC. This approach is used to characterize samples like coal derivatives [18], poly(dimethylsiloxane) (PDMS) [28], poly(methyl methacrylate) (PMMA) [29], polyester copolymers [27, 30], and a variety of synthetic polymers such as methacrylate copolymer, polybutylacrylate, polyester, polycarbonate, and polystyrene [33]. During the analysis of these samples, GPC offers narrow fractions to MALDI toward calibrating the GPC experiment. This offers a major improvement rather than creating narrow standards appropriate for calibration of GPC.
(b) Continuous collection of eluent (GPC fractions) on the right target prior to MALDI. This could be achieved by efficient spraying of the column eluent onto the target that contains the MALDI matrix. Equipment suitable for this includes the conventional liquid chromatography transform (LCT) from Lab Connections Inc. [14], robotic interface [32], or with home-built units [49]. This approach supports sample mass analysis, which is suitable for samples with broad polydispersity.
(c) Analysis of the GPC eluent by direct on-line MALDI [16, 41]. Fei and Murray [11] employed this technique for low molecular weight poly(ethylene glycol) (PEG) and poly(propylene glycol) (PPG) standards. The column eluent is channeled to the mass spectrometer as an aerosol, which contains both the analyte and the matrix.

3.2.3 X-ray Diffraction (XRD)

XRD analysis helps to provide important solid-state structural properties like degree of crystallinity. Polymers could be semicrystalline, highly crystalline, or amorphous microcrystalline, or a combination of the three forms. The composition of these forms depends on the polymer formulation and synthesis route, which in turn influences the mechanical properties such as creep, buckling, tensile strength, and compression. Hence, it is essential to determine the degree of crystallinity accurately [31]. The XRD profile of polysaccharides exhibits distinct peaks, two peaks at approximately $2\theta = 11.6°$ and $20.1°$ at (100) reflection [44] for chitosan. Chitosan has two different orthorhombic crystal forms with diffraction angles on the (100) plane. One of the orthorhombic (form I at the weak reflection peak 11.6°) exhibits a unit cell with $a = 7.76$ Å, $b = 10.91$ Å, and $c = 10.30$ Å. Form II has the strongest reflection peak 20.1°, which exhibits a unit cell with $a = 4.4$ Å, $b = 10.0$ Å, and $c = 10.3$ Å. XRD analysis is always conducted using with Cu Kα radiation ($\lambda = 1.5406$ Å). Figure 3.1 presents the XRD analysis of chitosan, polyelectrolyte complexes (PEC), and poly (methacrylic acid) (PMAA). When the polysaccharides are functionalized with polymers such as poly (2-methacryloyloxyethyl) trimethyl ammonium chloride (PDMC), they become amorphous due to the destruction of their ordered laminated structures [45, 46].

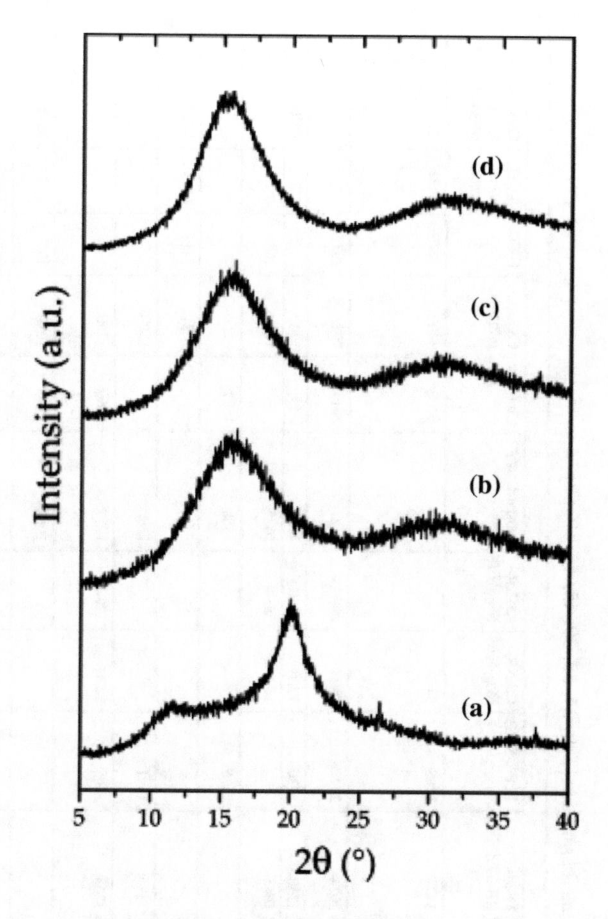

Fig. 3.1 X-ray diffractograms of chitosan (**a**), PEC [*r*COOH/NH_2 = 5.6 (**b**) and *r*COOH/NH_2 = 11.1 (**c**)], and PMAA (**d**). Figure is adapted from De Vasconcelos et al. [44]

3.2.4 Fourier Transform Infrared Spectroscopy (FTIR)

Polymeric adsorbents exhibit numerous functional groups, which are suitable for adsorption. FTIR analysis is used for confirming the presence of adsorption site in the polymeric adsorbent [34]. Several researchers have carried out FTIR analysis on polymeric adsorbent [35, 36]. Prominent adsorption bands responsible for adsorption are C–O–C groups at ~ 1273 cm^{-1} and C=O groups at ~ 1731 cm^{-1}. Furthermore, when the adsorbent is covalently bonded with graphene oxide (GO), a band at ~ 1636 cm^{-1}, which is assigned to –CONH– emerges for magnetic polymers modified (MP) with GO and functionalized with amine (NH_2-MP@GO) [48]. Table 3.1 presents FTIR vibration assignments of some selected polymeric adsorbent. Functionalization of magnetic polymers with amino group enhances their adsorption capacity toward heavy metal ions removal from polluted effluents. Shen et al. [39] studied the effect of incorporation of amino groups on nano-magnetic polymer adsorbents (NH_2-NMPs) toward adsorption of Cu(II) and

Table 3.1 Recorded FTIR vibrations assignments for polymeric samples [3, 9, 24, 35, 39, 45, 46, 50, 52]

Suggested assignment	Wave numbers (cm^{-1})														
	NH_2-MP@GO	MGO@MIP	NH_2-NMPs	Chitosan/sepiolite	OA-M	M-*co*-(GMAMMA)	TEPA-NMP	CS-NR-Mag-MIP	CS-NR-Mag-NIP	GO-Fe_3O_4@P	Chitosan	Chitosan-*g*-PDMC	β-CD	β-CD nano-sphere with	Cross-linked starch polymer
OH, NH stretching vibration and intermolecular hydrogen bonds for polysaccharide	3425	3402	3365	3600–3300	3446	3446	3365	3440		3400	3400	3400	3275	3400	3314
C H stretching vibration		2923	2924	2925	2924	2924	2924	2994	2990	2921	2911	2911	2925		
C=O groups	1731	1728	1730			1730	1730	1730	1730	1755		1729		1731	1720
C=O vibration of the amide group	1636			1655	1630	1645		1638	1638	1635		1630	1646		
Amine N–H vibration		1564	1576				1576			1568	1597				
Amine N–H vibration												1481	1413	1443	1450
N–H deformation vibrations				1392		1381	1381			1392		1390			
C–O–C stretching of secondary alcohol	1273		1265			1265	1265	1270	1270	1277	1260	1260			
C–O stretching of Primary alcohol				1080						1068	1080	1080	1077	1032	1025
Quaternary ammonium salt												954	998		
Characteristic peak for Fe_3O_4	580	589			589	589	589	563	563	594					

NH_2-MP@GO: Amine-functional magnetic polymer-modified graphene oxide, MGO@MIP: magnetic polymer-modified graphene oxide, NH_2-NMPs: amino-functionalized nano-Fe_3O_4 magnetic polymers, OA-M: oleic acid-coated Fe_3O_4 nanoparticles, M-*co*-(GMAMMA): glycidylmethacrylate and methyl methacrylate-functionalized nano-Fe_3O_4 magnetic polymers, TEPA-NMP: tetraethylenepentamine-functionalized nano-Fe_3O_4 magnetic polymers

CS-NR-Mag-MIP: bisphenol A-imprinted core–shell nano-ring amino-functionalized superparamagnetic polymer GO-Fe_3O_4@P: Surface core–shell magnetic polymer-modified graphene oxide, chitosan-*g*-PDMC: chitosan-grafted PDMC [poly (2-methacryloyloxyethyl) trimethyl ammonium chloride]

β-CD: Beta cyclodextrin

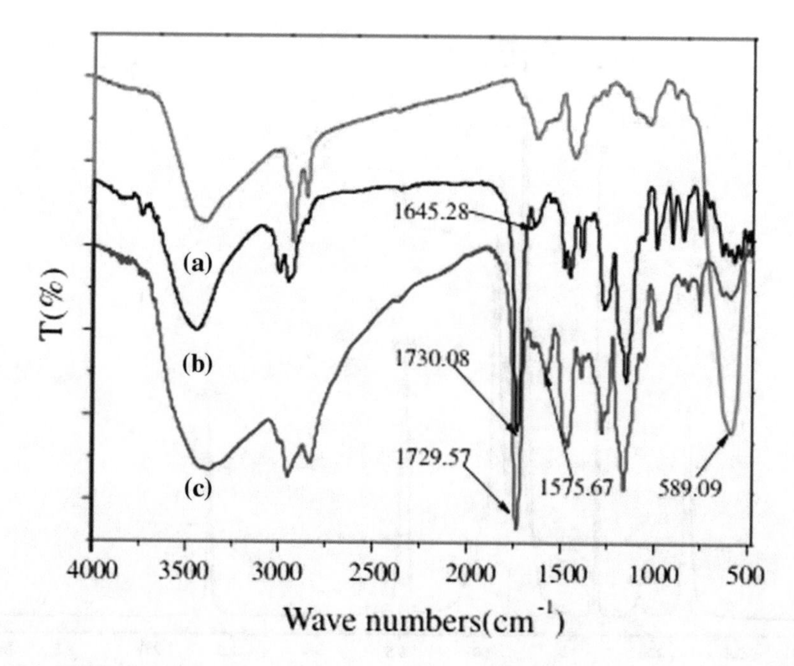

Fig. 3.2 FTIR adsorption spectra of **a** OA-M, **b** M-*co*-(GMAMMA), and **c** TEPA-NMPs [51]

Cr(VI) from water. Their result shows that the incorporated amino groups played a vital role in adsorbing the metal ions, and the higher the amount of amino groups in the sample, the higher the maximum adsorption capacity (q_m). [50] used glycidylmethacrylate (GMA) as a functional monomer in preparation of core–shell nano-ring magnetic molecularly imprinted polymer (MIP) functionalized with amino group. The incorporation of magnetic properties into GO synergistically combines the high adsorption capacity of GO with the convenience magnetic controllable separation as well as the high q_m of amine [36]. Marrakchi et al. [24] also reported that amino group containing polymeric adsorbent (chitosan) at acidic pH exhibits remarkable performance. The presence of hydroxyl and amino group in polymeric adsorbents gives them the capacity to adsorb a variety of contaminants such as metal ions, dyes, proteins, and phenolic compounds [4]. Figure 3.2 presents FTIR spectra for functionalized Fe_3O_4 magnetic polymers for removal of Cr(VI) from wastewater.

3.2.5 *Nuclear Magnetic Resonance (NMR)*

NMR is a viable technique for analysis, identification, and characterization of various kinds of polymers. It also provides information on the mobility of molecules to be characterized [9]. Chitosan is a derivatized biopolymer from chitin by

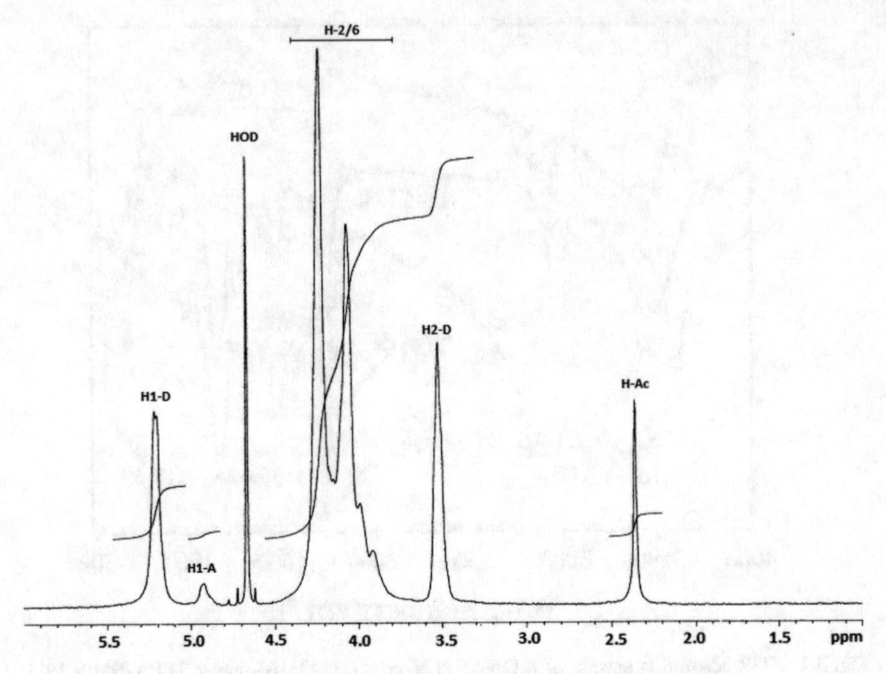

Fig. 3.3 Chitosan PCCH00005 (DDA ~ 87%) 1H NMR spectrum at 70 °C [21]

partial deacetylation. 1H NMR is used to determine the degree of deacetylation of chitin to form chitosan (Fig. 3.3), and the degree of deacetylation (DDA) could be computed as follow:

$$\mathrm{DDA}(\%) = \left(\frac{\text{H1-D}}{\text{H1-D} + \text{H-Ac}/3}\right) \times 100$$

where H1-D is the integrals of peak of proton H1 of deacetylated monomer, and H-Ac is the integrals of peak of proton H1 of the peak of the three protons of acetyl group.

NMR is also efficient in the determination of polysaccharides' degree of substitution [6] as they possess numerous reactive groups (acetamido and/or hydroxyl groups) situated at the 2-, 3-, and 6-positions in the glucose unit. The structure of modified or functionalized polymeric materials by grafting, cross-linking, etc., could also be confirmed by using C-solid-state NMR analysis. This indicates that NMR can confirm reactions like hydrolysis of the epoxy groups to diols [37]. However, static NMR is inadequate due to a number of interactions in polymeric adsorbents including chemical shift anisotropy and dipole–dipole interaction, and large heterogeneity, which resulted in redundant broad proton NMR spectra. Furthermore, analyzing with carbon–proton dipole interaction by using cross-polarization/magic-angle spinning with dipolar decoupling (CP/MAS) technique is excellent in the analysis of rigid structures, but exhibits poor performance as the sample mobility increases [5]. Consequently, 2D heteronuclear and homonuclear spectroscopy using high-resolution magic-angle spinning (HRMAS) is developed. In this new technique, the sample will be rotated at the magic angle, 54.7°, and HRMAS probe head will be used [25]. Martel et al. [25] used this technique to determine the chemical shift of β-cyclodextrin (CD)-linked chitosan derivative. They discovered about three resonances ranging from 4.8 to 5.7 ppm, which are ascribed to anomeric protons. This implies that there are a number of glucosidic derivatives (CD-substituted unit, glucosamine units, and acetylated units of the polymer chain) as well as the glycosidic and triazinyl-substituted glycosidic units of the grafted derivative.

Furthermore, solid-state HRMAS NMR with dipolar decoupling (DD-MAS) method with or without CP is efficient for both rigid and mobile phases [5, 8]. ^{13}C DD-MAS reveals signals arising mainly from the mobile component. This analysis is performed by varying the time between scan (D1), which correlates with the relaxation time T_1 to illustrate how rigid the sample is [8]. This was effectively demonstrated in the study of Crini et al. [5] when they used epichlorohydrin to cross-link cyclomaltoheptaose (-cyclodextrin). They observed different kinds of structure for the sample, polymerized epichlorohydrin and cross-linked β-CD, with different molecular mobility. The molar ratio of the sample under consideration is 20 (epichlorohydrin/CD) at 50 °C for 2 h using 50 wt% NaOH. Prior to the analysis of the cross-linked sample, they investigated the free β-CD sample. Table 3.2 presents the chemical shift data and the ^{13}C spin–lattice relaxation (T_1 in s) that determine the molecular mobility and homogeneity of ion exchangers obtained for the dry and hydrated states of β-CD and the polymeric adsorbent. The adsorbent is made up of 80 wt% epichlorohydrin, and its signals are overlapped by the C-2, -3, and -5 peaks of the parent β-CD. The signals in the region between 60 and 70 ppm increase (Fig. 3.4), while β-CD C-1 carbon signals decrease because of the hydroxymethyl group at C-6 in the glucose unit of β-CD or because of hydroxymethyl group of epichlorohydrin.

3.2.6 X-ray Photoelectron Spectroscopy (XPS)

XPS spectra could be used to determine the chemical composition of polymers. The compositions are identified by their binding energy, since each component and functional group has a specific binding energy. For instance, ester carbon (O–C=O), C–N, ether (C–O), and aliphatic carbon (C–C/C–H) have binding energies of 288.7, 288.2, 286.3, and 284.9 eV, respectively, in C1S XPS spectra. The XPS peaks at 532.1, 398.8, and 284.5 eV are attributed to oxygen (O 1s), nitrogen (N 1s), and carbon (C 1s), respectively, [50].

Table 3.2 ^{13}C spin–lattice relaxation (T_1 in s) of the carbon atoms of β-CD and β-CD-based polymeric adsorbent of both dry and hydrated (33% w/w of water) samples [5]

Carbon atoms of the polymer unit	Chemical shift (ppm)	β-CD		Polymeric adsorbent	
		Dry	Hydrated	Dry	Hydrated
		T_1	T_1	T_1	T_1
C-1	103.9	13.6			
C-2	102.9	20.9			
C-3	102.1	21.2			
	101.6	22.5	34		
	100.4			22	1.6
	99.6				
C-4	84.3	16.9			
	82.8	15.5			
	81.3	17.6	27.6	16.4	1.2
	78.0	18.5			
	74.7				
C-2	76.9	11.6			
C-3	76	10.8			
C-5	74.8	12.2			
	73.4	11.4			
	72.9				
	72.4	13.5	22.7	9.9	0.6
	71.9				
	70.6				0.6
	69.8				
C-6	63.9				
	62.5			nd[a]	0.4
	61.4				
	60.1	3.3			

[a]*nd* not determined

Recently, Mines et al. [26] employed XPS to verify the presence of polymer and its corresponding functional groups on the surface of activated carbon functionalized with covalent organic polymer. They observed a polymer shell attached to the carbon surface, which was validated by increase in the amount of the surface nitrogen from 0.00 to 32.80%. The nitrogen content is about one-third of the total grafted polymer. Moreover, a detailed observation of the C1s scan of the XPS measurement shows the presence of three separate peaks of C–N at 287.4, C=O at 286.3, and C at 284.3 eV, while the reference carbon shows only the plain C peak (Fig. 3.5).

3.3 Polymer Properties

3.3.1 Physical Properties

The remarkable sorption performance of polymeric materials such as polysaccharides is generally due to: (1) high hydrophilicity of the polymer, since they contain hydroxyl groups of glucose units, (2) the presence of numerous of functional groups such as hydroxyl, primary amino, and/or acetamido groups), (3) high chemical reactivity of these groups, and (4) flexibility of the polymer chain structure [4, 12]. However, starch-based materials are constrained due to their hydrophilicity. One of the notable polymeric materials is cyclodextrin, which is a product of this chemical derivation. It exhibits a remarkable physical property by forming inclusion compounds with several molecules, particularly aromatics. The interior cavity of the molecule offers a somewhat hydrophobic environment where polar contaminants can be trapped. On the other hand, some polymeric materials such as the styrene-divinylbenzene (SDB)-based (XAD-2 and XAD-4) are hydrophobic or nonpolar copolymer resin, which attracts hydrophobic organic matters as well as organic species that are sparingly soluble [10, 15]. Hydrophobic materials are attractive to compound like the phenolics from wastewater. Furthermore, polymethacrylate matrix-based materials such as XAD-7, which exhibits intermediate polarity, can also attract organic species [15]. Moreover, for cross-linked polymeric materials, the molecular attraction ability is limited to not only the hydrophilicity but also the cross-link density [40]. Delval et al. [8] developed cross-linked starch-based materials functionalized with amine group, which they used for removal of various dye from aqueous solution. They reported that the materials exhibit remarkable sorption capacities for acid, disperse, and reactive dyes, but are weak toward sorption of basic dyes. This remarkable performance is attributed to cross-link density of the materials.

The adsorption performance of polymeric materials is also influenced by the pH of the aqueous solution because the degree of dissociation, the surface charge, and the number of ionic species of contaminants [24] as well as the surface properties of the material depend on pH. Therefore, it is essential to consider the effect of pH of the solution toward understanding the link between the molecules of the contaminant and the surface of the adsorbent as the adsorption proceed.

3.3.2 Chemical Properties

Polymeric materials exhibit remarkable properties such as non-toxicity, high chemical reactivity, biocompatibility, chirality, adsorption capacities, chelation, and polyfunctionality. For instance, biopolymer like chitosan is soluble at acidic pH, thus cannot be used as an insoluble sorbent under these conditions. However, it

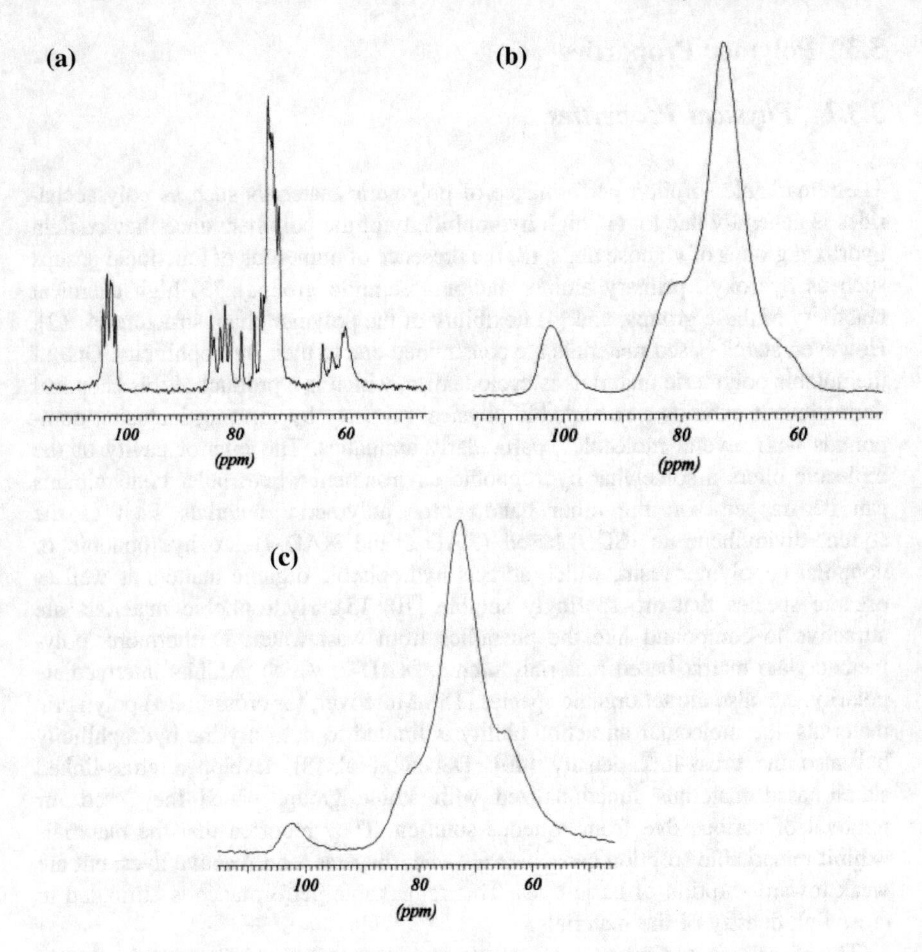

Fig. 3.4 **a** CP/MAS spectra of dry β-CD; **b** CP/MAS spectra of dry β-CD Polymer 4; **c** MAS spectra of dry β-CD Polymer 4 [5]

may become suitable after some physical and chemical modifications and functionalizations [4]. Amino-functionalized magnetic polymer is synthesized using Fe_3O_4 as the magnetic core exhibits a high chemical stability. The study of [50] reveals that the core–shell nano-ring magnetic MIP functionalized with amino group using glycidylmethacrylate (GMA) as a functional monomer is chemically stable. The stability was tested in acidic media with pH ranging from 0 to 3 for 12 h as well as in alkali media with pH ranging from 11 to 14. The result showed that synthesized polymeric material remains intact since the Fe_3O_4 did not leak out of the core, thereby retaining the magnetic property [50]. Meanwhile, the amino groups are easily protonated to predominantly form $–NH_4^+$ at lower pH (making it difficult to form hydrogen bond (–O–H···N) with the contaminant) and $–NH_2$ (forming hydrogen bonds (–O–H···N)). Deprotonation of hydroxyl group occurs at

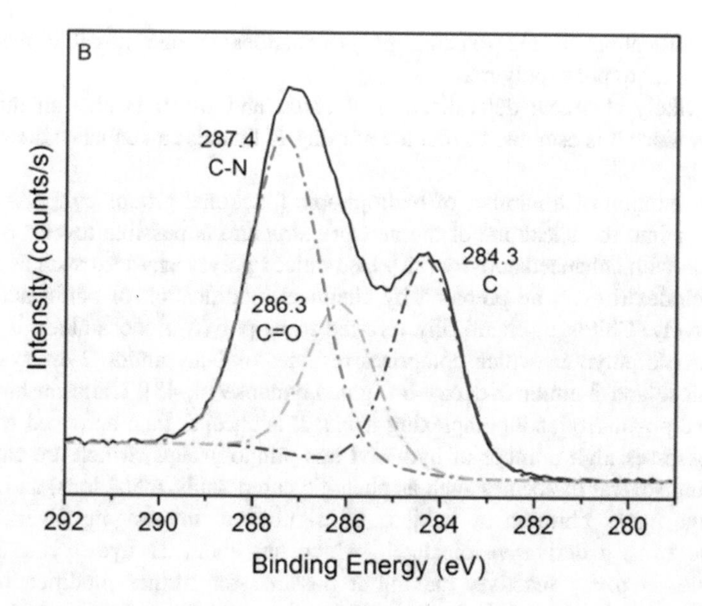

Fig. 3.5 XPS C1s scan of AC-COP (for interpretation of the references to color in this figure legend, the reader is referred to the Web version of the article [26])

pH above the optimum, making formation of hydrogen bond (–O–H···N) so difficult [35]. However, for amino-functionalized magnetic polymer-modified GO, a high adsorption capacity is possible at any pH because of the π–π stacking interaction between the GO sheet and the contaminant [3], and the carbonyl (–C=O) also could form hydrogen bond (–C=O···H) with the contaminant molecule easily [36].

Biopolymers like chitin and starch exhibit a good chemical reactivity due to their numerous reactive groups (acetamido and/or hydroxyl groups) situated at the 2-, 3-, and 6-positions in the glucose unit. With these groups, direct substitution reactions (etherification or esterification reactions) or chemical modifications (enzymatic degradation, oxidation or grafting reactions, hydrolysis), usually called chemical derivatization, to yield various derivatives of polysaccharide for particular domains of applications are imminent.

The derivatives of chitin and starch can be classified into three main classes of polymers [4]:

(i) modified polymers such as cationic starches, carboxymethylchitin,
(ii) derivatized biopolymers such as cyclodextrins, chitosan and their derivatives, and
(iii) polysaccharide-based materials like resins, membranes, gels, composite materials.

The chemical substitution reactions of chitin and starch have several challenges. The challenges are attributed to the presence of several functional groups, which take part in chemical reactions depending on the experimental conditions [19].

The modification of the existing polysaccharides is one possible method of obtaining more polar polymer.

The likely chemical derivatization of chitin and starch is also an interesting property since it is established that the grafting of ligands can enhance their activity [20, 42].

Incorporation of a number of hydrophobic functional groups either as pendant groups or into the backbone of the network structure is possible toward preparing materials with enhanced activity [38]. Derivatized polysaccharides such as chitosan and cyclodextrin can be prepared by chemical modification of chitin and starch, respectively. Chitosan, chemically referred to as poly(*N*-glucosamine), is a linear polycationic polymer, which comprises residues of 2-acetamido-2-deoxy-b-D-glucopyranose and 2-amino-2-deoxy-b-D-glucopyranose [4, 43]. Chitosan has gained immense popularity as a complexing agent; it is cheaper than activated carbon. It also possesses high number of hydroxyl and amino groups, which are capable of extracting several molecules such as phenolic compounds, metal ions, and dyes [1, 2]. This made chitosan a subject of significant interest in diverse fields. Cyclodextrins, a derivative of starch, which has about 21 hydroxyl groups per molecule, is partly reactive, making it possible for further modification using cross-linker such as epichlorohydrin (EPI) to improve its performance [7, 47].

3.4 Summary

Information on the physical, chemical, and characterization of polymeric materials is necessary to derive the relationship between their physicochemical and chemical properties. Such relations are vital toward rational design of economical but yet efficient polymeric materials. Particularly, we discussed GPC, XRD, Fourier transform infrared spectroscopy (FTIR), NMR, X-ray photoelectron spectroscopy (XPS), and physical and chemical properties of biopolymer-based and magnetic polymer-based materials, which individually probe a particular aspect of the materials. Therefore, a combination of these techniques is vital toward achieving a balanced description of the complexity of polymeric materials. Moreover, a proper understanding of the characteristics and properties of polymeric materials will enable researchers to design a suitable material for whatever industrial application they are required.

References and Future Readings

1. Babel S, Kurniawan TA (2003) Low-cost adsorbents for heavy metals uptake from contaminated water: a review. J Hazard Mater 97(1):219–243
2. Bailey SE, Olin TJ, Bricka RM, Adrian DD (1999) A review of potentially low-cost sorbents for heavy metals. Water Res 33(11):2469–2479

3. Chen M-L, Min J-Q, Pan S-D, Jin M-C (2014) Surface core–shell magnetic polymer modified graphene oxide-based material for 2, 4, 6-trichlorophenol removal. RSC Adv 4(108):63494–63501
4. Crini G (2005) Recent developments in polysaccharide-based materials used as adsorbents in wastewater treatment. Prog Polym Sci 30(1):38–70
5. Crini G, Cosentino C, Bertini S, Naggi A, Torri G, Vecchi C, Janus L, Morcellet M (1998) Solid state NMR spectroscopy study of molecular motion in cyclomaltoheptaose (β-cyclodextrin) crosslinked with epichlorohydrin. Carbohydr Res 308(1):37–45
6. Crini G, Martel B, Torri G (2008) Adsorption of CI Basic Blue 9 on chitosan-based materials. Int J Environ Pollut 34(1–4):451–465
7. Crini G, Morcellet M (2002) Synthesis and applications of adsorbents containing cyclodextrins. J Sep Sci 25(13):789–813
8. Delval F, Crini G, Bertini S, Filiatre C, Torri G (2005) Preparation, characterization and sorption properties of crosslinked starch-based exchangers. Carbohydr Polym 60(1):67–75
9. Delval F, Crini G, Morin N, Vebrel J, Bertini S, Torri G (2002) The sorption of several types of dye on crosslinked polysaccharides derivatives. Dyes Pigm 53(1):79–92
10. Euvrard É, Morin-Crini N, Druart C, Bugnet J, Martel B, Cosentino C, Moutarlier V, Crini G (2016) Cross-linked cyclodextrin-based material for treatment of metals and organic substances present in industrial discharge waters. Beilstein J Org Chem 12(1):1826–1838
11. Fei X, Murray KK (1996) On-line coupling of gel permeation chromatography with MALDI mass spectrometry. Anal Chem 68(20):3555–3560
12. Güven O, Şen M, Karadağ E, Saraydın D (1999) A review on the radiation synthesis of copolymeric hydrogels for adsorption and separation purposes. Radiat Phys Chem 56(4):381–386
13. Hanton S (2001) Mass spectrometry of polymers and polymer surfaces. Chem Rev 101 (2):527–570
14. Hanton SD, Liu XM (2000) GPC separation of polymer samples for MALDI analysis. Anal Chem 72(19):4550–4554
15. Hewitt LM, Marvin CH (2005) Analytical methods in environmental effects-directed investigations of effluents. Mutat Res/Rev Mutat Res 589(3):208–232
16. Hill SE, Feller D, Glendening ED (1998) Theoretical study of cation/ether complexes: alkali metal cations with 1, 2-dimethoxyethane and 12-crown-4. J Phys Chem A 102(21):3813–3819
17. Håkansson K, Zubarev RA, Håkansson P, Laiko V, Dodonov AF (2000) Design and performance of an electrospray ionization time-of-flight mass spectrometer. Rev Sci Instrum 71(1):36–41
18. Johnson BR, Bartle KD, Domin M, Herod AA, Kandiyoti R (1998) Absolute calibration of size exclusion chromatography for coal derivatives through MALDI-ms. Fuel 77(9–10): 933–945
19. Khor E, Lim LY (2003) Implantable applications of chitin and chitosan. Biomaterials 24 (13):2339–2349
20. Kumar MNR (2000) A review of chitin and chitosan applications. React Funct Polym 46 (1):1–27
21. Lavertu M, Xia Z, Serreqi A, Berrada M, Rodrigues A, Wang D, Buschmann M, Gupta A (2003) A validated 1 H NMR method for the determination of the degree of deacetylation of chitosan. J Pharm Biomed Anal 32(6):1149–1158
22. Li J, Du Y, Yang J, Feng T, Li A, Chen P (2005) Preparation and characterisation of low molecular weight chitosan and chito-oligomers by a commercial enzyme. Polym Degrad Stab 87(3):441–448
23. Lin H, Wang H, Xue C, Ye M (2002) Preparation of chitosan oligomers by immobilized papain. Enzyme Microb Tech 31(5):588–592
24. Marrakchi F, Khanday W, Asif M, Hameed B (2016) Cross-linked chitosan/sepiolite composite for the adsorption of methylene blue and reactive orange 16. Int J Biol Macromol 93:1231–1239

25. Martel B, Devassine M, Crini G, Weltrowski M, Bourdonneau M, Morcellet M (2001) Preparation and sorption properties of a β-cyclodextrin-linked chitosan derivative. J Polym Sci Part A Polym Chem 39(1):169–176
26. Mines PD, Thirion D, Uthuppu B, Hwang Y, Jakobsen MH, Andersen HR, Yavuz CT (2017) Covalent organic polymer functionalization of activated carbon surfaces through acyl chloride for environmental clean-up. Chem Eng J 309:766–771
27. Montaudo MS, Montaudo G (1999) Bivariate distribution in PMMA/PBA copolymers by combined SEC/NMR and SEC/MALDI measurements. Macromolecules 32(21):7015–7022
28. Montaudo G, Montaudo MS, Puglisi C, Samperi F (1995) Molecular weight distribution of poly (dimethylsiloxane) by combining matrix-assisted laser desorption/ionization time-of-flight mass spectrometry with gel-permeation chromatography fractionation. Rapid Commun Mass Spectrom 9(12):1158–1163
29. Montaudo MS, Puglisi C, Samperi F, Montaudo G (1998) Application of size exclusion chromatography matrix-assisted laser desorption/ionization time-of-flight to the determination of molecular masses in polydisperse polymers. Rapid Commun Mass Spectrom 12(9):519–528
30. Montaudo MS, Puglisi C, Samperi F, Montaudo G (1998) Molar mass distributions and hydrodynamic interactions in random copolyesters investigated by size exclusion chromatography/matrix-assisted laser desorption ionization. Macromolecules 31(12):3839–3845
31. Murthy NS (2004) Recent developments in polymer characterization using x-ray diffraction. Parameters 18:19
32. Nielen MW (1998) Polymer analysis by micro-scale size-exclusion chromatography/MALDI time-of-flight mass spectrometry with a robotic interface. Anal Chem 70(8):1563–1568
33. Nielen MW, Malucha S (1997) Characterization of polydisperse synthetic polymers by size-exclusion chromatography/matrix-assisted laser desorption/ionization time-of-flight mass spectrometry. Rapid Commun Mass Spectrom 11(11):1194–1204
34. Pan S-D, Chen X-H, Li X-P, Cai M-Q, Shen H-Y, Zhao Y-G, Jin M-C (2015) Double-sided magnetic molecularly imprinted polymer modified graphene oxide for highly efficient enrichment and fast detection of trace-level microcystins from large-volume water samples combined with liquid chromatography–tandem mass spectrometry. J Chromatogr A 1422: 1–12
35. Pan S-D, Shen H-Y, Zhou L-X, Chen X-H, Zhao Y-G, Cai M-Q, Jin M-C (2014) Controlled synthesis of pentachlorophenol-imprinted polymers on the surface of magnetic graphene oxide for highly selective adsorption. J Mater Chem A 2(37):15345–15356
36. Pan S-D, Zhou L-X, Zhao Y-G, Chen X-H, Shen H-Y, Cai M-Q, Jin M-C (2014) Amine-functional magnetic polymer modified graphene oxide as magnetic solid-phase extraction materials combined with liquid chromatography–tandem mass spectrometry for chlorophenols analysis in environmental water. J Chromatogr A 1362:34–42
37. Radi S, Ramdani A, Lekchiri Y, Morcellet M, Crini G, Janus L, Martel B (2000) Extraction of metal ions from water with tetrapyrazolic macrocycles bound to Merrifield resin and silica gel. J Appl Polym Sci 78(14):2495–2499
38. Shahidi F, Arachchi JKV, Jeon Y-J (1999) Food applications of chitin and chitosans. Trends Food Sci Technol 10(2):37–51
39. Shen H, Pan S, Zhang Y, Huang X, Gong H (2012) A new insight on the adsorption mechanism of amino-functionalized nano-Fe_3O_4 magnetic polymers in Cu(II), Cr(VI) co-existing water system. Chem Eng J 183:180–191
40. Shiftan D, Ravenelle F, Mateescu MA, Marchessault RH (2000) Change in the V/B polymorph ratio and T1 relaxation of epichlorohydrin crosslinked high amylose starch excipient. Starch-Stärke 52(6–7):186–195
41. Sutjianto A, Curtiss LA (1998) Li^+–diglyme complexes: barriers to lithium cation migration. J Phys Chem A 102(6):968–974
42. Synowiecki J, Ali Al-Khateeb N (2003) Production, properties, and some new applications of chitin and its derivatives. Crit Rev Food Sci Nut 43(2):145–171

43. Varma A, Deshpande S, Kennedy J (2004) Metal complexation by chitosan and its derivatives: a review. Carbohydr Polym 55(1):77–93
44. De Vasconcelos C, Bezerril P, Dantas T, Pereira M, Fonseca J (2007) Adsorption of bovine serum albumin on template-polymerized chitosan/poly (methacrylic acid) complexes. Langmuir 23(14):7687–7694
45. Wang J-P, Chen Y-Z, Yuan S-J, Sheng G-P, Yu H-Q (2009) Synthesis and characterization of a novel cationic chitosan-based flocculant with a high water-solubility for pulp mill wastewater treatment. Water Res 43(20):5267–5275
46. Wang J, Deng B, Wang X, Zheng J (2009) Adsorption of aqueous Hg (II) by sulfur-impregnated activated carbon. Environ Eng Sci 26(12):1693–1699
47. Xu W, Wang Y, Shen S, Li Y, Xia S, Zhang Y (1989) Studies on the polymerization of β-cyclodextrin with pichlorohydrin. Chin J Polym Sci 1:16–22
48. Yang X, Wang Y, Huang X, Ma Y, Huang Y, Yang R, Duan H, Chen Y (2011) Multi-functionalized graphene oxide based anticancer drug-carrier with dual-targeting function and pH-sensitivity. J Mater Chem 21(10):3448–3454
49. Yun H, Olesik SV, Marti EH (1999) Polymer characterization using packed capillary size exclusion and critical adsorption chromatography combined with MALDI-TOF mass spectrometry. J Microcolumn Sep 11(1):53–61
50. Zhao Y-G, Chen X-H, Pan S-D, Zhu H, Shen H-Y, Jin M-C (2013) Self-assembly of a surface bisphenol A-imprinted core–shell nanoring amino-functionalized superparamagnetic polymer. J Mater Chem A 1(38):11648–11658
51. Zhao Y-G, Shen H-Y, Pan S-D, Hu M-Q (2010) Synthesis, characterization and properties of ethylenediamine-functionalized Fe_3O_4 magnetic polymers for removal of Cr (VI) in wastewater. J Hazard Mater 182(1):295–302
52. Zhao Y-G, Shen H-Y, Pan S-D, Hu M-Q, Xia Q-H (2010) Preparation and characterization of amino-functionalized nano-Fe_3O_4 magnetic polymer adsorbents for removal of chromium (VI) ions. J Mater Sci 45(19):5291–5301

Chapter 4
Polymers for Coagulation and Flocculation in Water Treatment

Oladoja Nurudeen Abiola

Abstract The desires to improve on the operational efficiency of coagulation/flocculation (CF), a unit process in water and wastewater treatment, and to obviate the other challenges synonymous with the use of inorganic coagulants (i.e. aluminium- and iron-based alum) impelled the search for alternative coagulants that can ameliorate the identified shortcomings. Amongst the array of synthetic and natural origin materials that have been screened as alternatives to the conventional inorganic alum, polymeric coagulants have shown better promise. The inherent structural features of polymeric coagulants enhanced the CF process operation and economy. This treatise is an exposition on the variables that define the choice of polymeric coagulant as an alternative to the conventional inorganic coagulants. The theoretical bases for the choice of the different polymeric coagulants were discussed. Using the identified active coagulating species in the different polymeric coagulant as a premise, the underlying CF mechanisms in the use of this genre of coagulants were expounded. The research gap in the use of polymeric coagulant as substitute to the conventional inorganic coagulant was also highlighted.

4.1 Introduction

Coagulation/flocculation (CF) is a two-stage process that involves the destabilization and agglomeration of dispersed, dissolved, colloidal and suspended particles in the dispersion medium. In the conventional municipal wastewater treatment plants, the coagulation stage, where the coagulants are mixed into the water to be treated, is preceded by screening and other pre-treatment methods (Fig. 4.1).

Ordinarily, large suspended particles in an aqua system would settle without any form of treatment, if the system is stagnant and left undisturbed. In the case that the conditions required for the settling of these dispersed particulate matters are not

O. N. Abiola (✉)
Hydrochemistry Research Laboratory, Department of Chemical Sciences,
Adekunle Ajasin University, Akungba Akoko, Nigeria
e-mail: bioladoja@yahoo.com; nurudeen.oladoja@aaua.edu.ng

R. Das (ed.), *Polymeric Materials for Clean Water*,
Springer Series on Polymer and Composite Materials,
https://doi.org/10.1007/978-3-030-00743-0_4

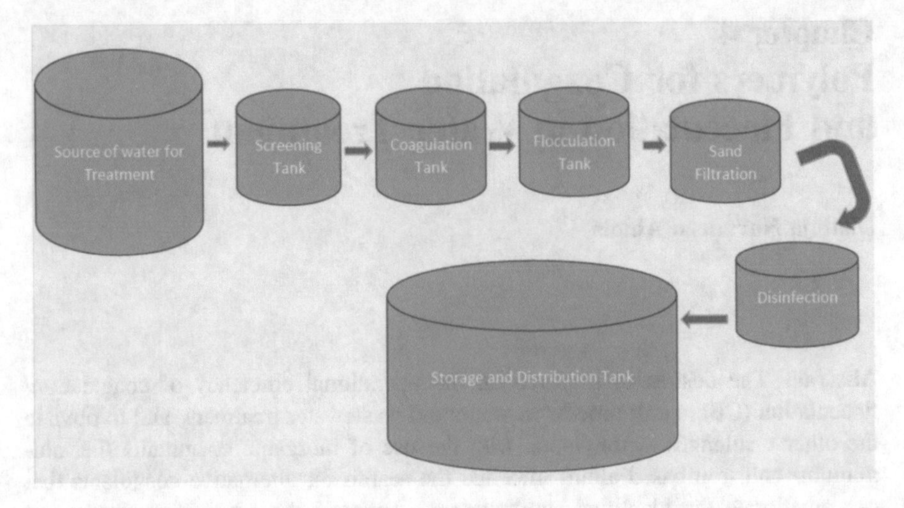

Fig. 4.1 Schematic representation of a typical conventional water treatment plants showing the different unit operations

met, the settling becomes unattainable. The settling of the dispersed particulates also becomes tasking if colloidal particles formed the large chunk of the dispersed phase in the aqua matrix. Consequently, water clarification using sedimentation alone is an unfeasible approach and a new functional approach, known as coagulation/flocculation, that enhanced the settling rate of the dispersed fractions of the aqueous system was developed. In the CF process, the stable dispersed particles are transformed to unstable and insoluble agglomerates, which are denser than water and then settle easily.

The process of CF involves series of chemical and mechanical operations that entails the addition and mixing of coagulant with the water matrix. These processes occur in two distinct stages that are termed as coagulation and flocculation. The first stage is the coagulation stage that involves rapid mixing of the coagulant with the water to be treated, which leads to the formation of micro-flocs, while the second stage (i.e. the flocculation stage) is the slow agitation stage that involves the aggregation of the micro-flocs to form the well-defined macro-flocs.

4.2 Polymeric Materials

Polymeric materials are made up of molecules which have long sequences of one or more species of atoms or groups of atoms linked to each other, usually by covalent bonds. The polymeric materials are also referred to as “macromolecules” and they are formed by the covalent linkage of small molecules, known as monomer, through a process known as polymerization. The long-chain skeletal framework of polymers sets them apart from other materials and gives rise to their characteristic properties.

The skeletal framework of polymers occurs in linear, branched and network forms. While linear skeletal structure can be represented by a chain with two ends, branched polymers have side chains, or branches, of significant length which are bonded to the main chain at branch points (also known as junction points), and are characterized in terms of the number and size of the branches. Network polymers have three-dimensional structures in which each chain is connected to all others by a sequence of junction points and other chains. Such polymers are said to be cross-linked and are characterized by their cross-linked density or the degree of cross-linking, which is related directly to the number of junction point per unit volume. Nonlinear polymers (i.e. branched and network polymers) are formed either by polymerization or prepared by linking together (i.e. cross-linking) of pre-existing polymer chains. Variations in skeletal structure give rise to major differences in properties.

4.3 Polymers in Coagulation–Flocculation

From the perspective of the chemical composition, the common coagulants used in CF operations can be broadly classified as organic and inorganic based. The inorganic-based coagulants include alum (aluminium sulphate), sodium aluminate, ferric sulphate, ferrous sulphate, ferric chloride, etc. The organic-based coagulants are mainly polymeric materials and can be further classified as synthetic- and natural-based coagulants. Synthetic chemical coagulants are based on organic polymers (or polyelectrolytes) while the natural-based organic coagulant is derived from both plants and animals.

In water and wastewater treatment operations, organic polyelectrolytes are used either for the CF operation or for sludge dewatering. Relative to the use of the inorganic coagulants in water treatment, the use of polymeric coagulants significantly reduced the coagulant dose requirements, the sludge volume, the ionic load cum the residual aluminium concentrations in the treated water and the process economy (i.e. cost savings of up to 25–30%) [33, 47]. The use of polymeric coagulants also enhanced the rate of floc settling and improved the floc textures in low-temperature coagulation or in soft coloured waters [14]. Since larger and stronger flocs are obtained from the use of organic polyelectrolytes, the capacity of a treatment facility and the rate of solid and water phase separation is substantially increased. In addition, the dosage of other ancillary chemicals required for water treatment is reduced and the coagulant is applicable in the treatment of broader range of water and wastewater.

Despite the numerous advantages of organic polyelectrolytes in CF operations, the use is also bedecked with challenges. Some of the derivatives of synthetic polymers are found to be non-biodegradable and the intermediate products of their degradation are hazardous to human health as their monomer is neurotoxic and carcinogenic [48]. Consequent upon the challenges synonymous with the use of synthetic polymeric materials in CF operations, the option of deriving polymeric coagulants from natural-based materials is now being explored. It was posited [51] that the polymeric coagulants derived from plants and animals are promising alternatives to the conventional coagulants because they are highly biodegradable, non-toxic and non-corrosive. They produced less voluminous sludge and do not alter the pH of the treated water. Furthermore, since the plants are locally grown, the natural plant-based coagulants are more cost-effective than imported chemicals [49].

4.3.1 Synthetic Polymeric Coagulants

The classification of the synthetic organic polymers used in water and wastewater treatment operations is based on the ionic charge present, e.g. cationic, anionic and non-ionic polyelectrolytes. Cationic polyelectrolytes formed the largest number of polymeric coagulants that have been used in CF operations. Bolto [3] provided a review of the varieties of cationic polymers used as coagulants in water and wastewater treatment operations. Most often, the cationic polyelectrolytes used possess quaternary ammonium groups that have a formal positive charge, irrespective of the pH value, and are referred to as strong polyelectrolytes [2]. Polymeric coagulants whose cationic properties are expressed only in acidic medium and considered weak polyelectrolyte are also used. Examples of cationic polyelectrolytes that have been used in CF operations include poly (diallyldimethyl ammonium chloride), epichlorohydrin/dimethylamine polymers, cationic polyacrylamides, etc.

Amongst the naturally occurring cationic polymers, chitosan is the most commonly used in CF operations. It is a partially deacetylated chitin and consists of a 1:4 random copolymers of *N*-acetyl-α-D-glucosamine and α-D-glucosamine. A review on the applications of chitosan in water industries was provided by Pariser and Lombardi [44]. Another natural cationic coagulant that is gaining popularity is the aqueous extract of the seed of *Moringa oleifera*, whose active coagulating ingredient has been identified as a cationic, low molecular mass peptide, whose isoelectric point of charge is above 10 [17, 18, 32]. Research reports [38, 50]; Jahn et al. [20, 30–32] have shown the performance efficiency of this seed extract in CF operations.

Relative to the synthetic polymers, polysaccharides are fairly shear stable and are biodegradable, but exhibit lower coagulation and flocculation efficiencies which always necessitate the use of higher dosage for optimal performance. Pal et al. [43] posited that in order to develop coagulants with better features than that of both the synthetic and natural polymeric coagulants, the best features of natural polymers can be combined with that of synthetic polymers via chemical modification. Naturally occurring non-ionic polymers have been transformed to cationic polymers through chemical modifications of the surface functional groups. A prominent example is the conversion of starch, a polymer composed of α-D-glucose units, to a cationic derivative by the reaction of the primary OH group with different cationic moieties. Incorporation of synthetic polymers via graft copolymerization reaction of synthetic polymers onto amylopectin, guar gum, glycogen and starch produced polymeric coagulants that are more effective than the original starting polymers because of the dangling polymer chains [2]. Kraft lignin, a coagulant with cationic character, has also been produced via the surface modification of lignin [27].

The commonly used anionic polymeric coagulant in water and wastewater treatment usually contains weakly acidic carboxylic acid groups; thus, the charge density is pH dependent [2]. A common example is anionic polyacrylamide. The anionic natural polymers are the sulphated polysaccharides or their derivatives, which include heparin, dextran sulphate, mannan sulphate, alginates, carrageenans, Gellan, gum Arabic and xanthan and chondroitin sulphate. These anionic polysaccharides found application more in medicine than in the water industries [59]. Few natural anionic polymers that have been tested in CF operations include tannins and modified natural polymer lignin sulphonate [28, 46].

Non-ionic polyacrylamide is a good example of the synthetic non-ionic polyelectrolytes while examples of natural non-ionic polysaccharides include amylose, amylopectin, cellulose, guar gum, etc. Natural, non-ionic polysaccharides that have been used in the water industries include starches, galactomannans, cellulose derivatives, microbial polysaccharides, gelatins and glues. Consequent upon the non-ionic nature of these polysaccharides, they are often used as coagulant aid instead of being used as the primary coagulant.

4.3.2 Natural Polymeric Coagulants

Aside the few natural polymeric coagulants, whose characteristics are well known, array of bio-based polymeric coagulants has also been derived from a range of biomaterials. It includes the fruit seeds of plant species, bone shell extracts, plant bark resins and extracts of the exoskeleton of shellfish (Fig. 4.2).

Choy et al. [7] provided a detailed review of the research report on twenty-one (21) types of plant-based coagulants derived from fruit waste and other biomaterials. The performance efficiencies of these green bio-based coagulants, the merits and demerits were enunciated. In another treatise, a review of fourteen (14) plant-based natural coagulants derived from common vegetables and legumes was provided [6]. The challenges of the different research efforts that focused on the use of bio-based coagulants were analysed as a prelude to further research efforts. Progress on the use of natural polymeric materials for water and wastewater purifications has also been chronicled by Oladoja [35]. Perspective on the promise and limitations of these bio-based polymeric coagulants were presented in this treatise. A synopsis of selected materials from which natural polymeric coagulants have been derived is presented in Table 4.1 [37].

Fig. 4.2 Photographic images of biomaterials from which natural polymeric coagulant have been derived (Plantago major fruit (**a**), Bean seeds of *Phaseolus vulgaris* (**b**), *Opuntia* Species (**c**), *Ipomoea quamocalit* (**d**), *Prosopis juliflora* (**e**), *Moringa Oleifera* Pod (**f**), Cassia obtusifolia plant (**g**), *Sterculia lychnophora* (**h**), Annona muricata (**i**))

4.4 The Underlying Mechanisms of the Process of CF

Premised on the classifications posited by Crittenden et al. [8], CF of the dispersed phase in an aqua system can be achieved via any or combination of the following operational mechanisms:

Table 4.1 Selected biomaterials studied as natural polymeric coagulants [37]

S/N	Scientific names	Common names	Family name	Country of origin	References
1	*Coccinia indica*	Ivy gourd, scarlet gourd, small gourd, kowai fruit, scarlet-fruited gourd	Cucurbitaceae	Central Africa, India and Asia	Lim [24], Shaheen et al. [52]
2	*Hibiscus esculentus*	Okra, lady's finger, gumbo, gobo	Malvaceae	Old World tropics (West Africa)	Small [54]
3	*Luffa cylindrica*	Smooth luffa, egyptian luffa, vegetable sponge, sponge guard	Cucurbitaceae	Old World tropics; probably Asia	Lim [24], Small [54]
4	*Arachis hypogaea*	Peanut, groundnut, monkey nut, pinder, goober	Fabaceae	South America	Lim [24], Boshou and Corley [4], Fageria et al. [13]
5	*Cicer arietinum*	Dal seeds, chick pea, bengal gram, garbanzo bean	Fabaceae	Mediterranean region	Lim [24], Ahmad et al. [1]
6	*Dolichos biflorus*	Horsegram, kulthi	*Fabaceae*	Old World tropics	Brink [5]
7	*Glycine max*	Soybean, soya bean	Fabaceae	Eastern Asia	Frederic [15]
8	*Guar gum*	Guar bean, cluster bean, guaran	Fabaceae	India	Peter et al. [45]
9	*Lablab purpureus*	Hyacinth bean, bonavist bean, chink, country bean, dolichos bean	Fabaceae	Old World tropics	Small [55]
10	*Phaseolus angularis*	Azuki bean, adsuki bean, red bean	Fabaceae	Unknown exact origin	Jansen [21]
11	*Phaseolus mungo*	Urad bean, black gram, black lentil, black matpe, urd bean	Fabaceae	India	Lim [24]
12	*Pisum sativum*	Green pea, pea, field pea, garden pea, stringless snow pea	Fabaceae	Southwestern Asia	Lim [24]

(continued)

Table 4.1 (continued)

S/N	Scientific names	Common names	Family name	Country of origin	References
13	*Vigna unguiculata*	Cow pea, black-eyed pea, southern pea, cowgram	Fabaceae	Southern Africa	Lim [24]
14	*Phaseolus vulgaris*	Common bean	Fabaceae	Central or South America	Fageria et al. [13]
15	*Cereus repandus*	Cadushi, giant club cactus, hedge cactus, peruvian apple cactus	Cactaceae	South America	Diaz et al. [9]
16	*Stenocereus griseus*	Pitaya agria, sour pitaya	Cactaceae	America	Fuentes et al. [16]
17	*Opuntia ficus-indica*	Prickly pears, tuna, nopal	Cactaceae	Americas	Zhang et al. [61] Miller et al. [29] Mane et al. [25] Shilpa et al. [53]
18	*Oryza sativa*	Rice	Poaceae	China	Thakre and Bhole [58]

(a) *Double-layer compression*

This mode of coagulation mechanism is initiated when a highly charged ionic species is used as the coagulant. This ionic coagulant tinkers with the overall ionic activity of the system and the electric double layer that encapsulates the dispersed particles is compressed and the repulsive energy barrier that keeps the particles apart is suppressed (Fig. 4.3a). In this case, molecular attraction is promoted and the formation of micro- and macro-flocs is enhanced.

(b) *Charge neutralization*

In this case, the oppositely charged ionic species derived from the coagulant is adsorbed onto the surface of the dispersed particulate matters. The negatively charged dispersed particles are neutralized by the positively charged coagulants and molecular attraction ensued (Fig. 4.3b).

(c) *Adsorption and bridging*

This mode of coagulation mechanism is common with polymeric coagulants. The skeletal framework of the polymeric coagulant induced the bridging of the dispersed particles, thereby bringing them together (Fig. 4.3c). Owing to the nature of the skeletal framework of polymeric coagulants, they are capable of extending

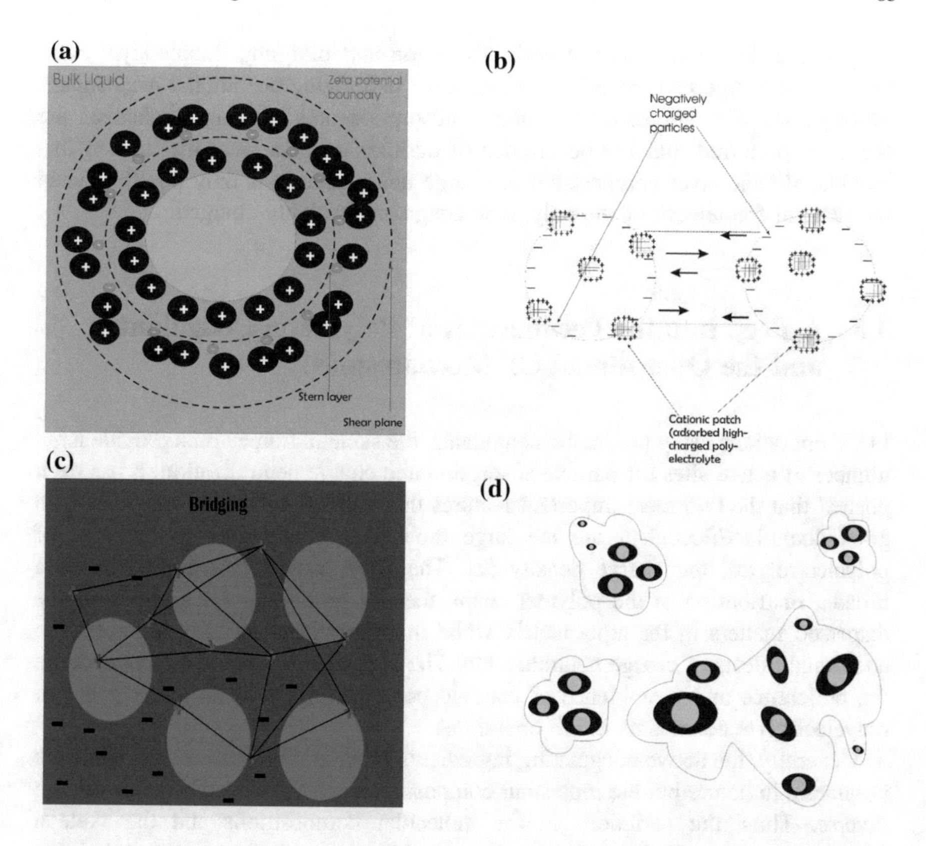

Fig. 4.3 **a** Schematic representation of double-layer compression mechanism, **b** schematic representation of charge neutralization mechanism, **c** schematic representation of adsorption and bridging mechanism, **d** schematic representation of sweep coagulation mechanism

into the solution to capture and bind multiple dispersed particles together to form denser and settleable flocs.

(d) *Sweep coagulation*

This mode of coagulation mechanism occurs through the enmeshment of the dispersed particulate matters into the matrix of the insoluble metal hydroxide formed from the hydrolysis of the metal coagulant (Fig. 4.3d). Relative to the charge neutralization mode of coagulation mechanism, sweep coagulation provides an improved coagulation and greater pollutant removal [12].

An overview of the underlying principles of the different coagulation mechanisms showed that the nature of coagulant influences the adopted operational coagulation mechanism. Alum coagulants (i.e. aluminium or ferric) can only operate via any or combination of sweep coagulation, charge neutralization and double-layer compression, but cannot operate using the adsorption and bridging mechanism. On the other hand, polymeric coagulants cannot operate using sweep coagulation, but it can operate with adsorption and bridging, double-layer compression or charge neutralization. Amongst the underlying operational mechanisms synonymous with polymeric coagulants, adsorption and bridging mechanism are the most preferred, but the occurrence of additional operational mechanism that includes double-layer compression or charge neutralization is only possible when the skeletal framework of the polymeric coagulant is highly charged.

4.5 A Peep into the Peculiarities of Polymeric Coagulants and the Operational CF Mechanism(S)

In CF operations using polymeric coagulants, the skeletal frameworks provide large number of active sites for particle adsorption and charge neutralization. It has been posited that the two most important features that defined polymeric coagulant as a good coagulant/flocculant are the large molecular weight and, in the case of polyelectrolytes, the charge density [2]. The large molecular weight ensures a broader distribution of the polymer chain, thereby promoting the bridging of the dispersed matters in the aqua matrix while the high charge density promotes the additional effects of charge neutralization. The high charge density effects underlie the preference and prominence of cationic polyelectrolytes to the other genre of polyelectrolyte coagulants in CF operations.

Generally, the active coagulating ingredients in natural bio-based coagulants are polymeric in nature but the molecular compositions and the skeletal framework are diverse. Thus, the variations in the molecular compositions and the skeletal framework are expected to influence the coagulation efficiencies and the underlying mechanisms of coagulation in different polymeric coagulants. The treatise [37] that provided mechanistic insights into the coagulation abilities of polysaccharide-based coagulants lend credence to the effects of the nature of the active coagulating ingredients on the CF abilities of polymeric coagulants.

The coagulation ability of chitosan, a linear copolymer of D-glucosamine and N-acetyl-D-glucosamine, was attributed to the high cationic charge density and ability to specifically bond to solid particles [37]. Owing to the high content of amine group on the polymer framework, it is positively charged at the pH value of natural water; thus, it effectively destabilizes and coagulates negatively charged suspended and colloidal matters and then flocculates them.

The seed gums obtained from diverse fruit seed (e.g. *Trigonella foenum-graecum*, *Cyamopsis tetragonoba*, *Ceratonia siliqua*, *Solanum lycopersicum*, *Coffea arabica*, *Convolvulaceae, Asteraceae, Arecaceae* etc.) have been studied as green bio-based polymeric coagulants. Thus far, the coagulating abilities of seed gums were attributed to either the presence of galactomannans or uronic acids. Since these two active coagulating ingredients are polymeric, the underlying coagulation mechanism by seed gums was ascribed to the bridging and adsorption mechanism, which is the coagulation mechanism specific to polymeric coagulants.

Galactomannans is a water soluble, macromolecular, hydrocolloid, having galactose and mannose in 1:2 molar ratio (Fig. 4.4a) while uronic acid (Fig. 4.4b) is a class of sugar acid with both carbonyl and carboxylic acid functional groups.

Different wastes from fruits, e.g. *Carica papaya, Feronia limonia, Mangifera indica, Persea americana, Phoenix dactylifera, Prunus armeniaca, Tamarindus indica*, have been screened as possible sources of natural polymeric coagulants and the coagulating abilities observed were attributed to the presence of long polymeric chain of unidentified natural polymers (i.e. proteins and polysaccharides) in the fruit wastes.

Mucilages derived from array of plants (e.g. *Opuntia ficus-indica* cactus, *Hibiscus esculentus, Plantago species, Malva sylvestris*) have been evaluated in CF operations. Mucilage, an hydrocolloid, is a complex polymer of carbohydrate with highly branched structure [26] that contains varying proportions of L-arabinose, D-galactose, L-rhamnose, and D-xylose, as well as galacturonic acid in different proportions. Premised on the chemical compositions of mucilage, the polysaccharide fraction was considered to be the active coagulating ingredient. This general assumption on the nature of the active ingredient present in mucilage was allayed by the findings of Miller et al. [29] who posited that there are fractional components of the *Opuntia* spp., outside those reported, that contributes to the coagulation activity.

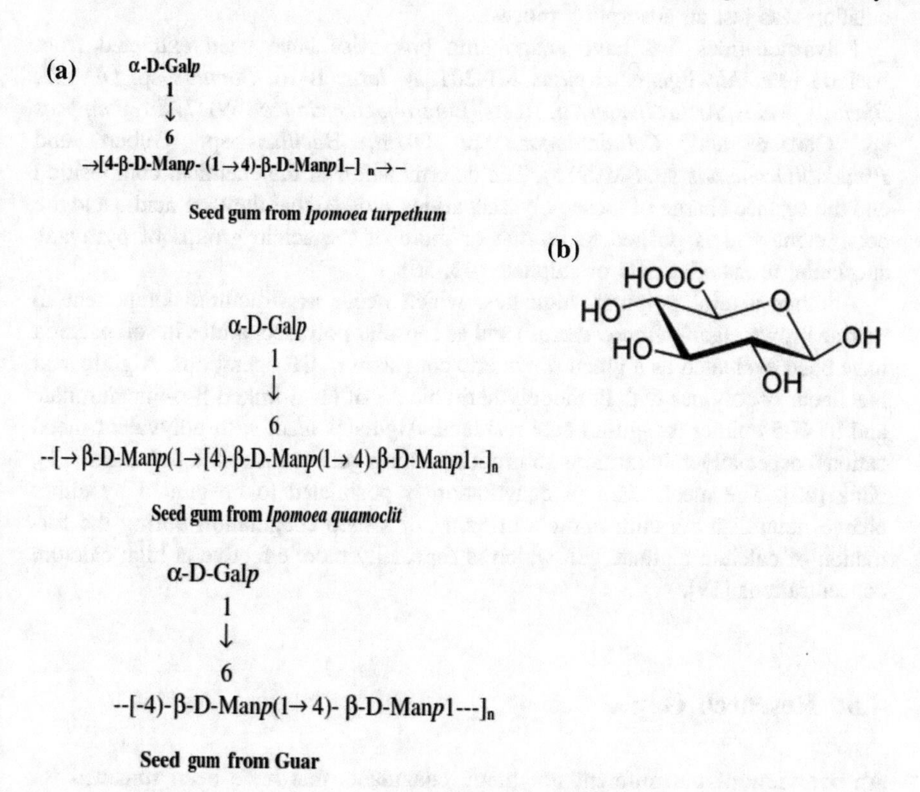

Fig. 4.4 **a** Illustration of the chemical structures of different galactomannans, **b** chemical structure of uronic acid (glucuronic acid)

The coagulating abilities of different fruit seeds (Nirmali, maize, mesquite bean, *Cactus latifaria*, *Cassia angustifolia*, leguminose, *Aesculus hyppocastanum*, *Quercus robur*, *Quercus cerris*, *Quercus rubra*, *Castanea sativa*, *Moringa Oleifera*, etc.) have been evaluated in CF operations. Premised on the position that the active coagulating species in *Moringa Oleifera* is a cationic protein, researchers have often concluded that the active coagulating species in other fruit seed coagulants are also proteins.

Tannins, a water-soluble polyphenolic compound, have been investigated as either the primary coagulant or coagulant aid in water treatment (Özacar and Sengil [40], [22, 36, 39, 41]. Polyphenols are characterized by the presence of manifold of phenol in the structural units. The anionic nature of tannin is caused by the phenolic groups, which is a good hydrogen donor. Consequent upon the anionic nature of tannins, they are used as natural anionic polyelectrolyte in CF operations.

The use of starch and starch-rich materials either as coagulant or coagulant aid [10], [34, 57] has been reported. Starch is a mixture of two polymers of anhydroglucose units (i.e. amylose and amylopectin) [60]. Albeit, there are reports on the coagulating ability of starch but Oladoja [35] disagreed with this assertion on the premise that the poor cationic charge density of starch precludes it from being a primary coagulant and that the process that was assumed to be coagulation–flocculation was just an adsorption process.

Polysaccharides that have coagulating properties have been extracted from bacteria (e.g. *Alcaligenes cupidus* KT-201, *A. latus* B-16, *Bacillus* sp. DP-152, *Bacillus firmus*, *Arthrobacter* sp. Raats, *Enterobacter cloacae* WD7, *Streptomyces* sp. Gansen and *Cellulomonas* sp. Okoh, *Bacillus* sp. Gilbert and *Pseudoalteromonas* sp. SM9913). The determination of the chemical composition and the surface charge of these polysaccharides showed that they are acidic and the component acid is posited to be one or more of the acidic groups of pyruvate, succinate, uronate, acetate or sulphate [42, 56].

Another natural polymer, alginates, which occur as structural component in marine brown algae (*phaeophyceae*) and as capsular polysaccharides in soil bacteria have been evaluated as a green polymeric coagulant in CF operations. Alginic acid is a linear copolymer with homopolymeric blocks of (1-4)-linked β-D-mannuronate and its C-5 epimer α-L-guluronate residues. Alginates react with polyvalent metal cations, especially calcium ions to produce strong gels or insoluble polymers [11], King [23]. The mechanism of coagulation is postulated to be guided by either charge neutralization cum particle bridging or sweep coagulation during the formation of calcium alginate gel, which is especially more effective at high calcium concentrations [19].

4.6 Research Gap

An overview of the different polymeric coagulants that have been screened for water and wastewater treatment showed the abundance of efforts that have been expended in this regard. Despite the retinue of laboratory reports that showed the comparable efficacy of the polymeric coagulants with the conventional inorganic coagulants, the use of the conventional inorganic coagulants, whose operational efficiency is laced with array of inadequacies are still in vogue.

Amongst the two classes of polymeric coagulants (i.e. synthetic- and natural-based coagulants), the use of synthetic coagulants is gradually becoming more acceptable in the water industries but the inherent challenges synonymous with the use are also a debilitating factor, despite the high operational efficiency that has been reported for its use. Considering the fact that the natural polymeric coagulants are highly biodegradable, non-toxic, non-corrosive, produces less voluminous sludge and does not alter the pH of the water under treatment, they have been found to be the promising alternative to the conventional inorganic coagulants. Despite the favourable features of the natural polymeric coagulants, the use has been restricted to few indigenous people in the under-resourced regions of the world.

In a comprehensive review [35] on headway on natural polymeric coagulants in water and wastewater treatment operations, detailed viewpoints on the research gap that accounted for the poor transition of laboratory findings to field trials and real-life applications in this field of research were chronicled. The synopses of these viewpoints, which gave an insight into the research gap that is militating against the real-life applications of natural polymeric coagulants in water and wastewater treatment, are presented below:

(a) The need to develop efficacious scientific strategies to isolate the active coagulating ingredient from the crude extracts to prevent the unusual overload of the treated water with organic matters.
(b) Development of proven scientific approach for the proper identification of the active coagulating component.
(c) Improper identification of the active coagulating species often leads to conflicting coagulation mechanisms proposed for coagulant derived from the same source
(d) Information on the overall quality characteristics of the treated water, which is pertinent to the evaluation of the overall safety of the treated water are rarely provided in reports from bench-scale studies.
(e) In most cases, the proposed coagulation mechanism by authors is based on speculations and data obtained from unrelated literature and not on any rigorous scientific expedition.
(f) The development of effective advocacy strategies for the dissemination of research findings to the appropriate governmental and non-governmental agencies, who served as an interface between the end-users and the researchers, is required.

4.7 Summary

The shortcomings identified with the use of the conventional inorganic coagulant were the impetus for the development of polymeric-based coagulant in water and wastewater treatment. The inherent skeletal framework of the polymeric coagulant that positioned them as efficacious coagulant is the large molecular weight and, in the case of polyelectrolytes, the charge density. Premised on the skeletal framework of polymeric coagulants, they cannot operate via the CF mechanism that bothered on sweep coagulation, but it can operate with mechanism that is governed by adsorption and bridging, double-layer compression or charge neutralization. Bridging the identified research gaps in this field of research would enhance the chances of getting research efforts in this field transposed from bench-scale to real-life applications.

References and Future Reading

1. Ahmad F, Gaur PM, Croser J (2005) Chickpea (*Cicer arietinum L.*). In: Singh RJ, Jauhar PP (eds) Genetic resources, chromosome engineering and crop improvement: grain Legumes. CRC Press, Boca Raton, FL, pp 187–217
2. Bolto B, Gregory J (2007) Organic polyelectrolytes in water treatment. Wat Res 41:2301–2324
3. Bolto BA (1995) Soluble polymers in water-purification. Progr Polym Sci 20(6):987–1041
4. Boshou L, Corley H (2006) Groundnut. CRC Press, Boca Raton, Florida
5. Brink M (2006) Macrotylomauniflorum (Lam.) Verde. In: Brink M, Belay G (eds), Plant resources of tropical Africa 1. Cereals and pulses, PROTA Foundation/Backhuys Publishers/CTA, Wageningen, 2006
6. Choy SY, Prasad KMN, Wu TY, Ramanan RN (2015) A review on common vegetables and legumes as promising plant-based natural coagulants in water clarification. Int J Environ Sci Technol 12(1):367–390
7. Choy SY, Prasad KMN, Wu TY, Raghunandan ME, Ramanan RN (2014) Utilization of plant-based natural coagulants as future alternatives towards sustainable water clarification. J Environ Sci 26:2178–2189
8. Crittenden JC, Trussell RR, Hand DW, Howe KJ, Tchobanoglous G (2005) Water treatment —principles and design, 2nd edn. Wiley, Hoboken, New Jersey
9. Diaz A, Rincon N, Escorihuela A, Fernandez N, Chacin E, Forster C (1999) A preliminary evaluation of turbidity removal by natural coagulants indigenous to Venezuela. Process Biochem 35:391–395
10. Dogu I, Arol AI (2004) Separation of dark-colored minerals from feldspar by selective flocculation using starch. Powder Technol 139:258–263
11. Draget KI, Smidsrod O, Skjak-Break G (2005) Alginates from algae, polysaccharides and polyamides in the food industry, properties, products and patents, in. Wiley, Weinheim, pp 1–30
12. Duan J, Gregory J (2003) Coagulation by hydrolysing metal salts. Adv Colloid Interfac Sci 100–102:475–502
13. Fageria NK, Baligar VC, Jones CA (2010) Growth and mineral nutrition of field crops. CRC Press, Boca Raton, Florida
14. Faust AD, Aly OM (1983) Chemistry of water treatment. Butterworths, Boston, pp 326–328
15. Frederic RJ (2004) The book of edible nuts. DoverPublications, USA
16. Fuentes SLC, Mendoza SIA, López MAM, Castro VMF, Urdaneta MCJ (2011) Effectiveness of a coagulant extractedfrom Stenocereusgriseus (Haw.) Buxb in water purification. RevTéc Ing Univ Zulia 34:48–56
17. Gassenschmidt U, Jany KK, Tauscher B, Niebergall H (1995) Isolation and characterization of a flocculation protein from *Moringa oleifera* Lam. BBA Biochem Biophys Acta 1243:477–481

18. Ghebremichael KA, Gunaratna KR, Henriksson H, Brumer H, Dalhammar G (2005) A simple purification and activity assay of the coagulant protein from *Moringa oleifera* seed. Water Res 39:2338–2344
19. Grant GT, Morris ER, Rees DA, Smith PJC, Thom D (1973) Biological interactions between polysaccharides and divalent cations: the egg-box model. FEBS Lett 32:195–198
20. Jahn SA, Musnad HA, Burgstalle H (1986) The tree that purifies water: cultivating multipurpose *moringaceae* in Sudan. Unasylva 38:23–28
21. Jansen PCM (2006) Vigna angularis (Willd.) Plant resources of tropical Africa 1. In: Brink M, Belay G (eds) Cereals and pulses. PROTA Foundation/Backhuys Publishers/CTA, Wageningen
22. Jeon JR, Kim EJ, Kim YM, Murugesan K, Kim JH, Chang YS (2009) Use of grapeseed and its natural polyphenol extracts as a natural organic coagulant for removal of cationic dyes. Chemosphere 77:1090–1098
23. King AH (ed) (1983) Brown seaweed extracts (alginates). In: Glicksman M (ed) Elsevier
24. Lim TK (2012) Edible medicinal and non-medicinal plants. Springer, New York
25. Mane PC, Bhosle AB, Jangam CM, Mukate SV (2011) Heavy metal removal from aqueous solution by Opuntia: a natural polyelectrolyte. J Nat Prod Plant Resour 1:75–80
26. Matsuhiro B, Lillo L, Saenz C, Urzu AC, Zarate O (2006) Chemical characterization of the mucilage from fruits of *Opuntia ficus indica*. Carbohydr Polym 63:263–267
27. McKague AB (1974) Flocculating agents derived from Kraft Lignin. J Appl Chem Biotechnol 24(10):607–615
28. Meister JJ, Li CT (1990) Cationic graft copolymers of lignin as sewage sludge dewatering agents. Polym Prepr Am Chem Soc, Div Polym Chem 31(1):664
29. Miller SM, Fugate EJ, Craver VO, Smith JA, Zimmerman JB (2008) Toward understanding the efficacy and mechanism of *Opuntia* spp. as a natural coagulant for potential application in water treatment. Environ Sci Technol 42:4274–4279
30. Muyibi SA, Evison LM (1996) Coagulation of Turbid water and softening of hardwater with *Moringa oleifera* Seeds. Int J Environ Stud 49:247–259
31. Muyibi SA, Okuofu CA (1995) Coagulation of low turbidity surface water with *Moringa oleifera* seeds. Int J Environ Stud 48:263–273
32. Ndabigengesere A, Narasiah KS, Talbot BG (1995) Active agents and mechanism of coagulation of turbid water using *Moringa oleifera*. Water Res 29:703–710
33. Nozaic DJ, Freese SD, Thompson P (2001) Long term experience in the use of polymeric coagulants at Umgeni Water. Water Sci Technol Water Supply 1(1):43–50
34. Oladoja NA (2014) Appraisal of cassava starch as coagulant aid in the alum coagulation of congo red from aqua system. Int J Environ Pollut Solut 2(1):47–58
35. Oladoja NA (2015) Headway on natural polymeric coagulants in water and wastewater treatment operations. J Water Process Eng 6:174–192
36. Oladoja NA, Alliu YB, Ofomaja AE, Unuabonah IE (2011) Synchronous attenuation of metal ions and colour in aqua stream using tannin–alum synergy. Desalination 271:34–40
37. Oladoja NA (2017) Mechanistic insight into the coagulation efficiency of polysaccharide based coagulants. In Oladoja NA, Unuabonah EI, Amuda OS, Kolawole OM (eds) Polysaccharides as green and sustainable resource for water and wastewater treatment. Springer Briefs in molecular science series
38. Olsen A (1987) Low technology water purification by bentonite clay and Moringa Oleifera seed flocculation as performed in sudanese villages: effects on *Schistosoma mansoni Cercariae*. Water Res 21:517–522
39. Özacar M, Sengil IA (2003) Enhancing phosphate removal from wastewater by using polyelectrolytes and clay injection. J Hazard Mater B 100:131–146
40. Özacar M, Sengil IA (2000) Effectiveness of tannins obtained from Valonia as a coagulant aid for dewatering of sludge. Water Res 34:1407–1412
41. Özacar M, Sengil IA (2003) Evaluation of tannin biopolymer as a coagulant aid for coagulation of colloidal particles. Colloids Surf A Physicochem Eng Asp 229:85–96

42. Pace GW, Righelato RC (eds) (1980) Production of extracellular microbial polysaccharides. Springer-Verlag, Berlin
43. Pal S, Mal D, Singh RP (2006) Synthesis, characterization and flocculation characteristics of cationic glycogen: a novel polymeric flocculant. Colloids Surf A—Physicochem Eng Asp 289 (1-3):193–199
44. Pariser ER, Lombardi DP (1989) Chitin sourcebook: a guide to the research literature. Wiley, New York
45. Peter RE, Qi W, Phillippa R, Yilong R, Simon R-M (2001) Guar gum: agricultural and botanical aspects, physicochemical and nutritional properties, and its use in the development of functional foods. In: Cho SS, Dreher ML (eds) Handbook of dietary fiber. Marcel Dekker Inc, New York
46. Rice DM, Denysschen JH, Stander GJ (1964) Evaluation of a tannin-base polyelectrolyte as a coagulant for turbid waters. CSIR Research Report 223, Pretoria
47. Rout D, Verma R, Agarwal SK (1999) Polyelectrolyte treatment an approach for water quality improvement. WaterSci Technol 40(2):137–141
48. Rudén C (2004) Acrylamide and cancer risk—expert risk assessments and the public debate. J Food Chem Toxicol 42:335–349
49. Sanghi R, Bhattacharya B, Dixit A, Singh V (2006) Ipomoea dasysperma seed gum: an effective natural coagulant for the decolorization of textile dye solutions. J Environ Manage 81:36–41
50. Schulz CR, Okun DA (1983) Treating surface waters for communities in developing countries. JAWWA 75(5):212–223
51. Sciban M, Klašnja M, Antov M, Škrbić B (2009) Removal of water turbidity by natural coagulants obtained from chestnut and acorn. Bioresour Technol 100:6639–6643
52. Shaheen SZ, Bolla K, Vasu K Singara, Charya MA (2009) Antimicrobial activity of the fruit extracts of Cocciniaindica. Afr J Biotechnol 8:7073–7076
53. Shilpaa BS, Akankshaa K, Girish P (2012) Evaluation of cactus and hyacinth bean peels as natural coagulants. Int J Chem Environ Eng 3:187–191
54. Small E (2011) Top 100 exotic food plants. CRC Press, Boca Raton, Florida
55. Small E (2009) Top 100 food plants. NRC Research Press
56. Sutherland IW (1977) Bacterial exopolysaccharides—their nature and production. Academic Press, London
57. Teh CY, Wu TY, Juan JC (2014) Optimization of agro-industrial wastewater treatment using unmodified rice starch as a natural coagulant. Ind Crops Prod 56:17–26
58. Thakre VB, Bhole AG (1985) Relative evaluation of a few natural coagulants. J Water Supply Res Technol 44:89–92
59. Voycheck CL, Tan JS (1993) Ion-containing polymers and their biological interactions. In: Hara M (ed) Polyelectrolytes: science and Technology. Dekker, New York, pp 309–310
60. Wei Y, Cheng F, Zheng H (2008) Synthesis and floculating properties of cationic starch derivatives. Carbohydrate Polymers 74: 673-679
61. Zhang JD, Zhang F, Luo YH, Yang H (2006) A preliminary study on cactus as coagulant in water treatment. Process Biochem 41:730–733

Chapter 5
Polymer and Polymer-Based Nanocomposite Adsorbents for Water Treatment

Bingcai Pan, Xiaolin Zhang, Zhao Jiang, Zhixian Li, Quanxing Zhang and Jinlong Chen

Abstract In the past decades, polymer and polymer-based nanocomposite adsorbents have been emerging as promising materials for the removal of various pollutants from contaminated waters, in terms of strong mechanical strength, excellent hydraulics performance, high stability, and tunable surface chemistry. In general, the adsorption of target pollutant is highly dependent upon the physicochemical structure of adsorbent materials, such as skeleton chemistry, pore structure, surface functional groups as well as the encapsulated moieties. This chapter reviews the synthesis, structure, and adsorption mechanism of polymer and polymer-based nanocomposite adsorbents utilized for the removal of various organic and inorganic pollutants. Also, the application of these materials is particularly concerned.

5.1 Introduction

Adsorption is listed among the most effective and simplest approaches to removing and recycling toxic pollutants from water/wastewater. A great amount of materials including activated carbon [1–3], cellulose [4], alginate [5, 6], and diatomite [7, 8] have been employed to achieve this purpose. Among these adsorbents, activated carbon might represent the most attractive and popular one to trap a variety of traditional pollutants such as phenols, dyes, and organic acids [9–13]. However, several intrinsic drawbacks of activated carbon, such as costly regeneration, high attrition rate, and indiscriminate adsorption toward organic pollutants, raise an imperative requirement to develop other more efficient adsorbents for water treatment [14–18]. Generally, in order to achieve wider application, the new adsorbents

B. Pan (✉) · X. Zhang · Z. Jiang · Z. Li · Q. Zhang · J. Chen
State Key Laboratory of Pollution Control and Resource Reuse,
School of the Environment, Nanjing University, Nanjing 210023, China
e-mail: bcpan@nju.edu.cn

B. Pan · X. Zhang · Q. Zhang
Research Center for Environmental Nanotechnology (ReCENT),
Nanjing University, Nanjing 210023, China

R. Das (ed.), *Polymeric Materials for Clean Water*,
Springer Series on Polymer and Composite Materials,
https://doi.org/10.1007/978-3-030-00743-0_5

should exhibit some promising properties like high adsorption capacities, strong mechanical strength, easy operation, and satisfied reusability.

Since the birth of synthetic ion exchanger, polymer adsorbents of cross-linked nature have been emerging as promising adsorbents for over seven decades due to their high surface area, satisfied mechanical strength, tunable surface chemistry, and pore size distribution [19]. Target pollutants in water can be sequestrated by polymer adsorbents via various mechanisms including electrostatic attraction [20], pore filling [21, 22], hydrogen bond formation [23, 24], π–π interactions [21], and hydrophobic interaction [25]. By utilizing well-designed polymer adsorbents, organic pollutants like aromatic or polyaromatic hydrocarbons [26, 27], alkanes, and their derivatives [28] can be efficiently removed from aqueous systems. Particularly, bifunctional polymers were developed by modifying hyper-cross-linked resin with charged groups like amine groups, rendering the resultant adsorbents extremely efficient to sequestrate highly water-soluble organic pollutants including aromatic sulfonates [20, 29], aromatic acids [30–32], and phenolic compounds [30, 33]. Unlike activated carbon, the exhausted polymer adsorbents can be steadily refreshed for cyclic use under mild conditions [34]. In addition, nanotechnology has opened a door for rational design of highly efficient water purifiers. Inorganic nanoparticles (NPs) present as zero-valent metals, metal oxides, and metal phosphates exhibit preferable adsorption toward heavy metals, arsenic, phosphate, and fluoride arising from large surface-to-volume ratio and high reactivity [35]. However, ultrafine nature of NPs intensively hindered their application in scaled-up water treatment, because NPs faced some challenges such as tendency to aggregate, difficulty in operation, and possible risk when released into the environment [36]. In the past decade, a variety of polymer-based nanocomposites were developed by encapsulating inorganic nanoparticles (NPs) inside porous polymer hosts to incorporate high reactivity of NPs and easy operation of bulky polymer hosts [37–43].

This chapter focuses on the advance in polymer and polymer-based nanocomposite adsorbents utilized as water purifiers, discussing their structure, reactivity and selectivity, reusability, adsorption mechanism, as well as their application in practical water treatment. Also, the forthcoming development in the field was forecasted.

5.2 Polymer Adsorbent for Organic Pollutants

Typically, organic pollutants are captured by polymer adsorbents resulting from micropore filling, hydrophobic interaction, electrostatic attraction, hydrogen bonding, and even complex formation [27, 44]. Nonpolar polymer adsorbent has emerged half a century before and nowadays, a large number of polymeric adsorbents have been developed for the removal various organic pollutants from aqueous solutions [30, 32, 33, 45, 46]. To further improve the adsorption capacity and reactivity, hyper-cross-linked polymer adsorbents possessing abundant micropore and macropore simultaneously were developed [47]. Also, polyacrylic ester

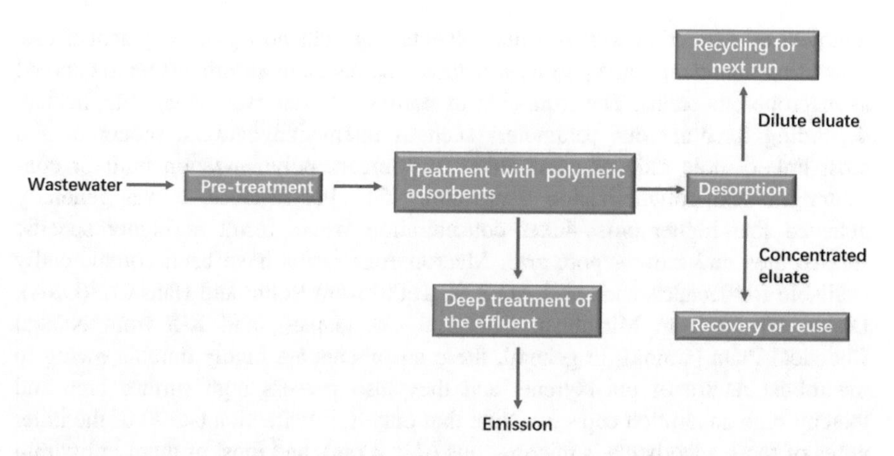

Fig. 5.1 A typical flowchart of industrial wastewater treatment by polymeric adsorbents. Redrawn from [50]

resins were developed in order to obtain new polymer adsorbents with improved hydrophilicity [48]. More recently, bifunctional polymer adsorbents were synthesized by chemical modification of traditional polymer adsorbents to realize specific adsorption toward some highly water-soluble organic pollutants [49].

As for the operation mode of polymer adsorbent, fixed-bed process is frequently adopted for their scaled-up application in chemical wastewater treatment, as illustrated in Fig. 5.1. In brief, after simple pretreatment like filtration, wastewater containing different pollutants undergoes adsorption with polymer adsorbents packed in the bed. Sometimes, the effluents require further treatment such as biochemical process, advanced oxidation process, and membrane filtration to meet stringent discharge standard. The loaded organic matters can be desorbed from the exhausted adsorbents during regeneration process, where organic solvents, acid, or alkaline solution are widely employed. Usually, the resultant dilute eluate is recycled as regenerant for next run, and the concentrated eluate can be recycled for recovery and reuse [47, 50].

5.2.1 Nonpolar Polymer for Hydrophobic Organic Pollutant

Due to nonpolar and hydrophobic characteristics, polystyrene resins were widely used for the removal of hydrophobic organic pollutants [28]. The polystyrene resin adsorbents can be obtained via free radical polymerization reactions using styrene, divinylbenzene and peroxysulfate or 2,2′-Azobis(2-methylpropionitrile) (AIBN) as the monomer, cross-linker, and initiator, respectively [19, 51], and the synthesis reaction often occurs in the presence of porogen like toluene [52], xylene [53], and linear polystyrene. The residual monomer, cross-linker, or porogen would be removed by extraction with organic solvents (e.g., ethanol) prior to practical use, generating abundant macroporous structure. The resultant adsorbents were denoted as macroporous resins. The formation of porous structure was rather complicated, depending on numerous parameters such as interaction between monomer and cross-linker, mole ratio of cross-linker to porogen, polymerization initiator concentration, and polymerization temperature [54]. For instance, it was generally believed that higher cross-linker concentration would result in higher specific surface area and narrow pore size. Macroporous resins have been commercially available for decades, such as XAD-2 (4, 1600) from Rohm and Haas Co. (USA), Diaion HP-20 from Mitsubishi Chemical Co. (Japan), and X-5 from Nankai Chemical Plant (China). In general, these adsorbents are highly durable owing to the robust nature of polystyrene, and they also possess high surface area and thereby high adsorption capacity. Note that only a little fraction (<5%) of the inner pores of these adsorbents is microporous ($d < 2$ nm), and most of them fall within the scope of macroporous ($d > 50$ nm) and mesoporous ($2 < d < 50$ nm) ranges. For instance, one of the most widely used adsorbent, Amberlite XAD-4, possesses ~ 880 m^2/g of BET surface area and its average pore diameter is ~ 11 nm [47]. The small amount of microporous structure suggests that micropore filling plays insignificant role in adsorption of organic pollutants onto these adsorbents, though high adsorption capacity still can be achieved due to high surface area. Adsorption affinity between pollutants and macroporous resin may be somewhat weak in the absence of specific interactions. Consequently, continuous efforts have been made to enrich microporous structure of these adsorbents to enhance their adsorption performance, as discussed below. In addition, polyacrylic ester resins are also utilized for hydrophobic pollutants sequestration. Table 5.1 summarizes representative examples using nonpolar polymers for decontamination of aqueous solutions from hydrophobic organic pollutants.

Unlike activated carbon, polymer adsorbents can be tailor-made and the adsorption process is often reversible; i.e., the adsorbate could be efficiently desorbed under mild conditions for resource recovery or further treatment. Such property renders nonpolar polymers extremely attractive for resource recycling of highly concentrated organic wastewater discharged from industries. For nonpolar polymer adsorbents, organic solvents such as methanol, alcohol, and acetone are among the most commonly utilized regeneration agents in terms of their strong dissolving ability toward a variety of hydrophobic matters.

5.2.2 Polar Polymer for Hydrophilic Organic Pollutant

Table 5.2 summarizes the recent literatures reporting polystyrene or polyacrylic ester adsorbents employed for the removal of hydrophilic organic matters (such as phenol, aniline, and chlorophenol) from water. Clearly, most polystyrene resins exhibit relatively poor adsorption toward hydrophilic pollutants in comparison with polyacrylic ester ones, attributing to polar nature of the latter. Polyacrylic ester

Table 5.1 Nonpolar polymer adsorbents used for the removal of hydrophobic organic pollutants from waters

Adsorbent	Type	Pollutant	Capacity	References
XAD-2	Polystyrene	Benzene	730 mg/g	[55]
XAD-4	Polystyrene	Benzene	1400 mg/g	[55]
XAD-2, 4	Polystyrene	Chlorobenzene	825 mg/g	[55]
XAD-4	Polystyrene	Chlorobenzene	2060 mg/g	[55]
NDA-16	Polystyrene	Chloronitrobenzene	325 mg/g	[56]
XAD-1600	Polystyrene	Dichloromethane	27 mmol/g	[28]
XAD-2	Polystyrene	Carbon tetrachloride	1250 mg/g	[55]
XAD-4	Polystyrene	Carbon tetrachloride	2600 mg/g	[55]
XAD-2	Polystyrene	Di-2-pyridyl ketone salicyloyl hydrazone	2.2 mg/g	[57]
XAD-4	Polystyrene	Diethyl phthalate	649 mg/g	[58]
HCP-1.3	Hyper-cross-linked	Benzene	30.0 mmol/g	[59]
NDA-150	Hyper-cross-linked	Diethyl phthalate	825 mg/g	[60]
XAD-7	Polyacrylic ester	Dichloromethane	17.8 mmol/g	[28]
XAD-7	Polyacrylic ester	Di-2-pyridyl ketone salicyloyl hydrazone	10.4 mg/g	[57]
XAD-7	Polyacrylic ester	Diethyl phthalate	480 mg/g	[58]

resins can be available via free radical polymerization reactions, which are similar to polystyrene ones, except that acrylate instead of styrene is utilized as the monomer, and their chemical structure is illustrated in Fig. 5.2 [48]. Acrylate is more reactive than styrene during polymerization reaction, resulting in superior mechanical stability of polyacrylic ester resin over polystyrene ones [61]. Polyacrylic ester adsorbents have been successfully applied for the removal of highly water-soluble compounds from water/wastewater, typical of which are reactive dyes and some sulfonated compounds (Table 5.2). Besides, the exhausted polyacrylic ester adsorbents can be regenerated more easily than polystyrene adsorbents. For instance, the 2-naphthalene sulfonate loaded polyacrylic ester resin (NDA-801) could be fully refreshed after simple wash by hot water (348 K) [48]. By using the process depicted in Fig. 5.1, polyacrylic ester adsorbents are capable of decontaminating highly concentrated organic effluents and recycling valuable hydrophilic organic matters. Nowadays, lots of polyacrylic ester adsorbents are commercially available, including Amberlite XAD-7, -7HP, and -8 from Rohm and Haas (US), Wofatit EP62 and Y59 from Chemie AG Bitterfeld (Germany), and NDA-7 from Jiangsu NJU Environ. Co. (China).

The adsorption of hydrophilic compounds by polyacrylic ester resin involves multiple interactions including hydrophobic interaction, electrostatic attraction, hydrogen-bonding formation, and even chemical adsorption depending on solution chemistry [44]. It is well recognized that hydrophobic interaction plays a significant

Table 5.2 Polystyrene or polyacrylic ester adsorbents used for hydrophilic organic removal from water

Adsorbents	Type	Pollutant	Capacity	References
XAD-2, 4	Polystyrene	Phenol	0.4–2.5 mmol/g	[62]
XAD-4	Polystyrene	phenol	0.60 mmol/g	[63]
XAD-4	Polystyrene	*p*-Cresol	1.18 mmol/g	[63]
XAD-4	Polystyrene	*p*-Chlorophenol	1.43 mmol/g	[63]
XAD-4	Polystyrene	*p*-Nitrophenol	1.20 mmol/g	[63]
XAD-4, 16	Polystyrene	Phenol	Not available	[64]
XAD-16	Polystyrene	Phenol	1.50 mmol/g	[33]
XAD-2, 4	Polystyrene	Phenol, salicylic acid	0.2–3 mmol/g	[32]
XAD-4	Polystyrene	Aniline	0.72 mmol/g	[65]
XAD-4	Polystyrene	Phenol	0.52 mmol/g	[65]
XAD-4	Polystyrene	Phenol	1.22 mmol/g	[66]
XAD-4	Polystyrene	4-Chlorophenol	1.47 mmol/g	[66]
XAD-2	Polystyrene	1-Naphthol	125.3 mg/g	[26]
XAD-2	Polystyrene	2-Naphthol	109.3 mg/g	[26]
XAD-2	Polystyrene	1-Naphthylamine	180.2 mg/g	[26]
XAD-2	Polystyrene	2-Naphthylamine	208.6 mg/g	[26]
XAD-4	Polystyrene	1-Naphthol	307.4 mg/g	[26]
XAD-4	Polystyrene	2-Naphthol	320.4 mg/g	[26]
XAD-4	Polystyrene	1-Naphthylamine	473.6 mg/g	[26]
XAD-4	Polystyrene	2-Naphthylamine	479.7 mg/g	[26]
XAD-4, 12, 16	Polystyrene	Benzoic acid	Not available	[67]
Duolite ES-861	Polystyrene	*m*-Cresol	141 mg/g	[23]
XAD-4	Polystyrene	4-Chlorophenol	30.89 mg/g	[25]
XAD-4	Polystyrene	Sodium 6-dodecyl benzene sulfonate	1.95 mmol/g	[68]
XAD-4	Polystyrene	Reactive brilliant blue KN-R	0.4 mmol/g	[68]
XAD-4	Polystyrene	Caffeine	28.5 mmol/g	[69]
XAD-4	Polystyrene	Cephalosporin C	23.4 mmol/g	[69]
XAD-7	Polyacrylic ester	Phenol	78.7 mg/g	[53]
XAD-7	Polyacrylic ester	4-Chlorophenol	1.31 mmol/g	[66]
XAD-8	Polyacrylic ester	Phenol	Not available	[67]
XAD-7	Polyacrylic ester	1-Naphenol	278.0 mg/g	[26]
XAD-8	Polyacrylic ester	1-Naphenol	296.5 mg/g	[26]
XAD-7, 8	Polyacrylic ester	1-Naphthylamine	256.4 mg/g	[26]
XAD-7, 8	Polyacrylic ester	1-Naphthylamine	267.1 mg/g	[26]
XAD-7, 8	Polyacrylic ester	2-Naphthylamine	232.5 mg/g	[26]
XAD-7, 8	Polyacrylic ester	2-Naphthylamine	263.2 mg/g	[26]

(continued)

Table 5.2 (continued)

Adsorbents	Type	Pollutant	Capacity	References
XAD-7	Polyacrylic ester	Linalool	Not available	[70]
XAD-7	Polyacrylic ester	Caffeine	58.3 mmol/g	[69]
NDA-801	Polyacrylic ester	Sodium 2-naphthalene sulfonate	123 mg/g	[48]

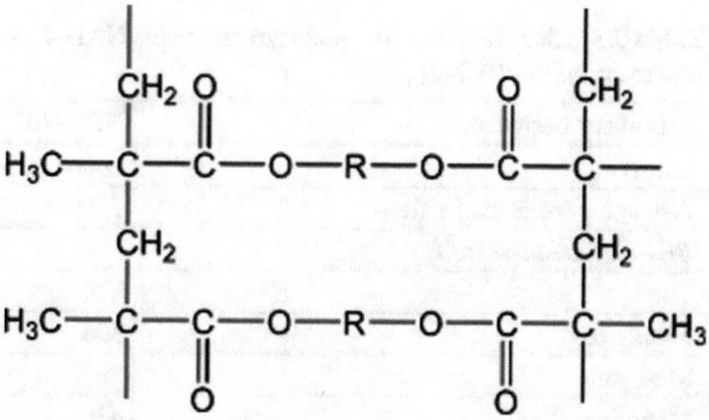

Fig. 5.2 Schematic structure of polyacrylic ester adsorbent (R denotes the cross-linking reagent, e.g., dimethyl acrylate glycol ester) [48]

role in adsorption [65, 71]. Besides, electrostatic attraction occurs between the positively charged ester groups and anionic pollutants under acidic pHs. Furthermore, recent study [48] elucidated that hydrophobic interaction and electrostatic attraction play a synergetic role in efficient removal of aromatic sulfonates by NDA-801.

5.2.3 *Advanced Polymer Adsorbents*

5.2.3.1 Hyper-Cross-Linked Polymers

To further improve the adsorption performance of polymer adsorbents, another polystyrene adsorbent, namely hyper-cross-linked resin, was developed at the end of the 1960s [72, 73]. Hyper-cross-linked resins are now commercially available, including NDA-701 and NDA-150 from NJU Environ. Co. (China), and Hypersol–Macronet MN-200 and MN-250 from Purolite (UK). They can be synthesized through Friedel–Crafts reaction on linear polystyrene [72] or post-cross-linking reaction on chloromethylated styrene-divinylbenzene (St-DVB) copolymer beads [47], and the latter one is more commonly used, possibly arising from its promising characteristics such as easy operation and the morphology of spherical beads. In brief, the preparation starts with suspension polymerization of styrene and divinylbenzene, followed by chloromethylation and Friedel–Crafts reaction on the resultant polystyrene beads [47]. For comparison, brief synthetic procedures of

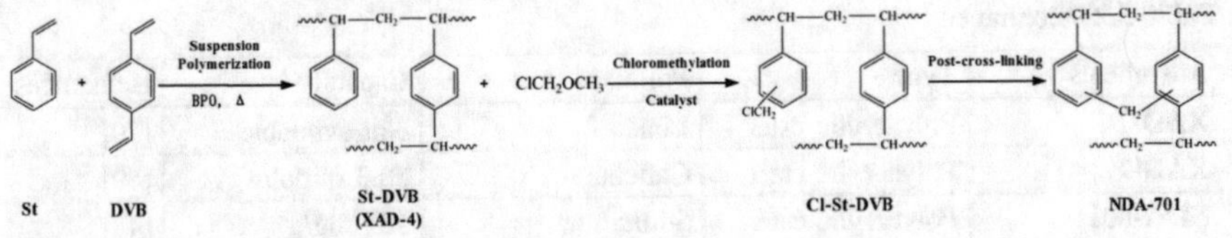

Fig. 5.3 Schematic illustration of synthetic procedures for a polystyrene adsorbent (XAD-4) and a hyper-cross-linked polymer adsorbent (NDA-701) [47]

Table 5.3 Characteristics of polystyrene resin XAD-4 and hyper-cross-linked polystyrene adsorbent NDA-701 [47]

Adsorbent designation	NDA-701	XAD-4
Matrix	Polystyrene	Polystyrene
Average pore diameter (nm)	2.24	5.61
BET surface area (m^2/g)		
Total	824	886
Macropore	306	196
Mesopore	88	651
Micropore	430	31
Pore volume (cm^3/g)	0.58	1.22
Micropore volume (cm^3/g)	0.22	0.0043
Density (wet, g/mL)	1.08	1.04
Particle size (mm)	0.5–1.0	0.5–1.0
Swelling ratio in benzene (%)	<5	>15
Osmotic-attrited perfect ball ratio (%)	>99.5	~92

XAD-4 and NDA-701 are illustrated in Fig. 5.3 [47], and the basis properties of both samples are listed in Table 5.3. Scanning electron micrographs of NDA-701 and XAD-4 are depicted in Fig. 5.4. Additionally, their pore size distribution is illustrated in Fig. 5.5 [47]. Clearly, NDA-701 possesses much more microporous structure over XAD-4, which was believed to greatly enhance adsorption toward organic pollutants through micropore filling, similar to activated carbon. Thus, hyper-cross-linked adsorbents often exhibit much higher adsorption capacity toward organic pollutants than macroporous polystyrene resins [47]. Moreover, considerable amount of macroporous structure (>200 nm) in NDA-701 facilitates fast diffusion of target pollutants inside the polystyrene beads, rendering NDA-701 with much faster adsorption kinetic than activated carbon [47]. The 4-nitrophenol-loaded NDA-701 is amenable to an entire regeneration by using NaOH solution as regenerant, whereas only ~75% regeneration efficiency was observed for activated carbon [47]. However, excessive cross-linking degree may be detrimental to the diffusion of target pollutant into the bulk polymer and consequentially compromise their adsorption capacity.

Fig. 5.4 Scanning electron micrographs of NDA-701 and XAD-4 [47]

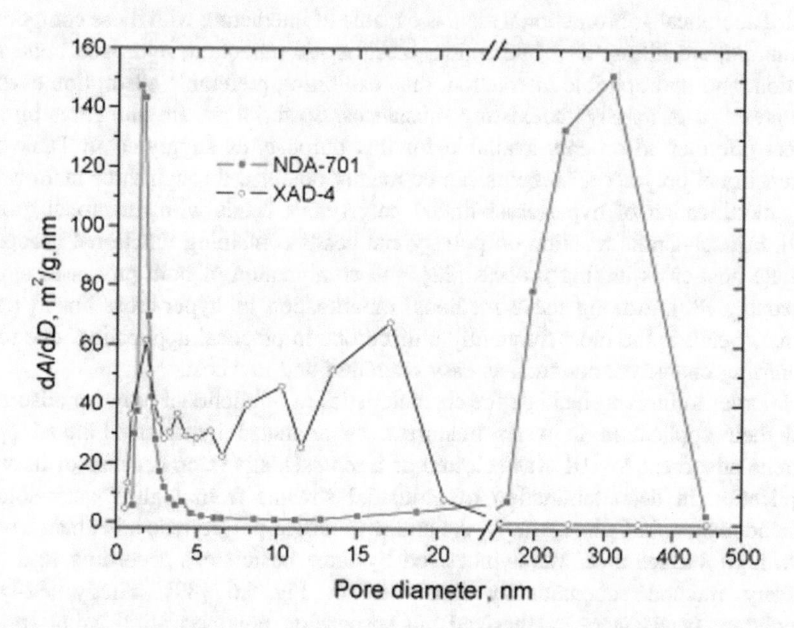

Fig. 5.5 Pore size distribution of polystyrene adsorbents NDA-701 and XAD-4 [47]

In addition to polystyrene materials, hyper-cross-linked polymer beads also include post-cross-linked polyacrylic adsorbent [74], the adsorbent using *p*-xylene-aromatics as skeleton [75], and 4,4′-bis(chloromethyl)-1,1′-biphenyl and benzene/biphenyl as skeleton [76]. Similarly, these hyper-cross-linked polymer adsorbents were mainly developed for improved adsorption toward target organic pollutants by taking advantage of increased microporous structure.

5.2.3.2 Bifunctional Polymer Adsorbents for Water-Soluble Organic Compounds

Considerable amount of organic pollutants in industrial effluents are highly water soluble, and they often contain charged groups including sulfonic group, amine group, carboxyl group, and hydroxyl groups. Traditional nonpolar polymer adsorbents exhibit poor adsorption toward these pollutants, due to weak interaction between adsorbent and adsorbate. In order to improve the adsorption capability of polymer adsorbents toward highly water-soluble organic compounds, bifunctional resin has been developed by modifying hyper-cross-linked resin with functional groups like amine, carboxyl, phenol, and sulfonic groups. For instance, many aromatic compounds like aromatic carboxylic acids, naphthalene, and benzene sulfonic acids, and quaternary benzylammonium compounds mainly exist as ions in water over a wide range of pH [77, 78]. Due to high solubility in water, they cannot be effectively removed by traditional nonpolar polymer adsorbents from aqueous system. Theoretically, bifunctional resin is capable of interacting with these compounds via micropore filling, π–π interaction, electrostatic attraction, hydrogen bond formation, and hydrophobic interaction, thus exhibiting preferable adsorption even in the presence of massive coexisting substances. So far, there are numerous bifunctional polymer adsorbents available for this purpose, as suggested in Table 5.4. Bifunctional polymer adsorbents can be mainly obtained through three main ways, i.e., modification of hyper-cross-linked polystyrene beads with functional groups [79], Friedel–Crafts reaction on polystyrene beads containing functional groups to initiate post-cross-linking process [22], and combination of both processes simultaneously [80]. Among these methods, modification of hyper-cross-linked polystyrene beads is the most frequently utilized one in practical application, due to its promising characteristics such as easy operation and low cost.

In order to further elucidate the characteristics of bifunctional polymer adsorbent and their application in water treatment, an aminated hyper-cross-linked polystyrene adsorbent M-101 was selected as a representative one because of its wide application in decontamination of industrial streams from highly water-soluble organic matters in China. Briefly, the amine-modified polystyrene adsorbent M-101 with high surface area was synthesized by three basic steps according to a proprietary method schematically illustrated in Fig. 5.6 [89]. Firstly, St-DVB copolymer beads were synthesized via suspension polymerization using styrene as monomer and divinylbenzene as cross-linker. Secondly, St-DVB were chloromethylated to create possible bond bridge for the following amination and the chloromethylated beads (Cl-St-DVB) were obtained. Thirdly, Cl-St-DVB were moderately post-cross-linked by Friedel–Crafts reaction, followed by amination with dimethylamine to obtain the resultant adsorbent M-101 [90].

Salient properties of the bifunctional adsorbent M-101 are listed in Table 5.5 [20], suggesting that M-101 possesses characteristics of both hyper-cross-linked polystyrene (e.g., high surface area and micropore volume) and anion exchanger (abundant covalently bonded amine groups). Extensive studies demonstrated M-101 as an excellent adsorbent for many ionic organic compounds like sulfonated

Table 5.4 Bifunctional polystyrene adsorbents for the removal of highly water-soluble organic pollutants from water

Functional group	Pollutant	Capacity	References
Tertiary amine	Phenol, benzoic acid, *o*-phthalic acid, benzene sulfonic acid, 2-naphthalenesulfonic acid	90–270 mg/g	[81]
Amidocyanogen	Phenol, aniline	Not available	[82]
Tertiary amino group	Sodium benzene sulfonate	1.2 mmol/g	[20]
Carboxyl group	p-Nitroaniline	3.2 mmol/g	[29]
Tertiary amino group	Resorcinol, catechol	1.0–1.8 mmol/g	[21]
Tertiary amine	Methomyl	40 mg/g	[83]
Sulfonic acid group	Methomyl	5 mg/g	[83]
Dicyandiamide	Reactive brilliant blue KN-R	28.1 mg/g	[84]
Sulfonic group	Acetylaminophenol	428.5 mg/g	[85]
Amino group	Naphthalene sulfonates	108 mg/g	[86]
Polyethylene glycol group	Yellow 5GL	Not available	[87]
2-carboxy-3/4-nitrobenzoyl, 2,4-dicarboxybenzoyl	Oxamyl, methomyl, desisopropylatrazine, phenol, dimethoate, atrazine, 2,4-dichlorophenoxy acetic acid, hydroquinone, resorcinol, catechol, orcinol, guaiacol	Not available	[49]
Lateral alkyl quaternary ammonium group	Cholate, taurocholate, chenodeoxycholate	1.35–1.73 mmol/g	[88]
1,2-dichloroethane	Bisphenol A	326.8 mg/g	[80]
Multiple phenolic hydroxyl groups	p-nitroaniline	Not available	[24]

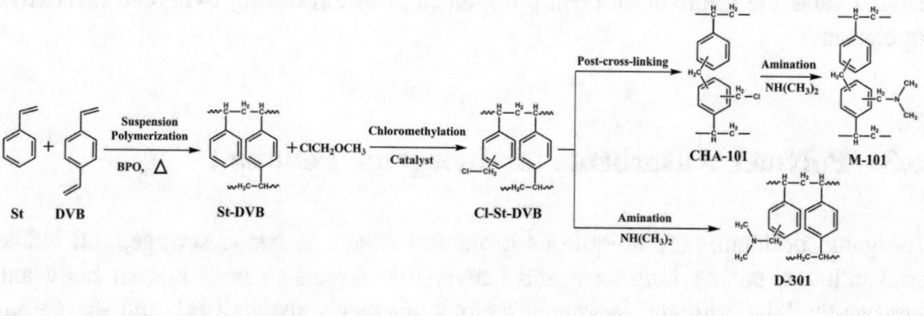

Fig. 5.6 Schematic illustration of synthetic procedures for an aminated polystyrene adsorbent D-301 and a hyper-cross-linked aminated polystyrene adsorbent M-101 [20]

Table 5.5 Salient properties of the aminated polystyrene adsorbent M-101

Cross-link density (%)	>35
BET surface area (m^2/g)	671.5
Macropore volume (cm^3/g)	0.16
Mesopore volume (cm^3/g)	0.028
Micropore volume (cm^3/g)	0.40
Total anion exchange capacity (mmol/g)	1.53
Quaternary ammonium group (mmol/g)	0.027

aromatics [20]. For instance, M-101 exhibits higher adsorption capacity toward sulfonated aromatics (e.g., benzene sulfonate, naphthalene sulfonate) than a hyper-cross-linked polystyrene adsorbent CHA-101 or a weakly basic anion exchanger D-301 [20]. Such surprising property of M-101 mainly results from π to π interaction between the aromatic structure of the polymeric matrix and nonpolar moiety of sulfonated aromatics, as well as electrostatic attraction between the positively charged amine groups and the negatively charged sulfo groups [90]. Note that M-101 has been successfully utilized as adsorbent to remove 1- and 2-naphthalene sulfonates (NS) from 2-naphthol manufacturing effluent (500 m^3/d, Chuanqing Chemical Plant, China) containing about 1500 mg/L 1-NS, 5000 mg/L 2-NS, and 7–12% of sodium sulfate [91]. After treatment, total NS in the effluent was reduced to <40 mg/L. Regeneration efficiency of the exhausted adsorbents could achieve 99% by using 2 M NaOH as regenerant, and the concentrated eluate after regeneration can be recycled to the production line for NS recovery. M-101-based adsorption technique has been employed in at least four plants in China, processing $\sim$400,000 m^3 industrial effluent annually. Another example is using M-101 adsorbent to treat the manufacturing wastewater of 4,4-dinitrobenzyl ethylene-2,2-disulfonic acid (i.e., DSD acid) in Huaihua Chemical Co. (Jiangsu, China) and Huayu Chemical Group (Hebei, China) [92]. Field applications suggested that DNS acid in industrial stream was reduced from $\sim$3500 to <30 mg/L by M-101, and the removal efficiency remained constant during two-year successive operation.

5.3 Polymer Adsorbents for Inorganic Pollutant

Inorganic pollutants are ubiquitous in natural water, domestic sewage, and industrial effluent, posing long-term and irreversible threats to both human body and ecosystem. The common inorganic cations are heavy metals [93], and the anions include arsenic [94], fluoride [95], nitrate [96], and phosphate [43, 97–99]. Theoretically, the inorganic ions can be removed by ion exchange and complex formation. Accordingly, various types of adsorbents are developed. Ion exchange adsorbents capture ionic pollutants through electrostatic attraction. Polymer chelating adsorbents were tailor-made to adsorb heavy metals through the complexation between heavy metals and chelating groups covalently bonded on the polymer matrix. Considering that many toxic pollutants can form inner-sphere complex with metal (hydr)oxides, polymer-based nanocomposites were designed by embedding metal (hydr)oxide nanoparticles (NPs) inside porous polymer matrix to overcome the intrinsic drawbacks of NPs such as tendency to aggregate together, difficult operation, and potential risk if released into environment. Moreover, multi-functional polymer nanocomposite adsorbents integrating with other functions like size exclusion and oxidation were developed to realize enhanced removal under rather complex medium [100, 101].

5.3.1 Polymers for Cationic Adsorption

5.3.1.1 Polymeric Cation Exchanger

Polymeric ion exchangers usually consist of three parts, i.e., insoluble polymer matrix, functional groups bound on the matrix, and replaceable charged counterions [102]. Ion exchange provides advantages in the removal of cations, including water softening and removal of heavy metal cations from water [103]. Typical cation exchange resin is usually available by modifying polystyrene skeleton with acidic functional groups. Commercially available cation exchange resins can be roughly divided into two types, strongly acidic type and weakly acidic type. Strongly acidic cation exchange resins are usually bound with sulfonic group ($-SO_3H$) and exhibit strong acidity like sulfuric acid, such as FPC11 (14 or 22) Na, FPC22 (23) H from Rohm and Haas Co. (USA), D001 from Zhengguang Resin Co. (China), DOWEX 50 W from Dow Chem. Co. (USA). Weakly acidic cation exchange resins are usually modified with carboxyl group (-COOH) and exhibit weak acidity like organic acid, such as FPC3500 from Rohm and Haas Co. (USA) and D113 from Zhengguang Resin Co. (China). Strong acid cation exchange resins are effective in hard water softening [104] and the removal of some heavy metal ions [103] such as zinc, nickel, or copper from acid solution.

However, ion exchange mainly involves electrostatic attraction mechanism, thus exhibiting nonselective adsorption toward cations in water. As a result, the removal of heavy metals was intensively suppressed by coexisting mineral cations at greater levels, like Na^+, K^+, Ca^{2+}, and Mg^{2+} [105]. Many efforts have been made to improve adsorption selectivity of polymer adsorbents toward heavy metals, particularly in the presence of massive coexisting cations.

5.3.1.2 Polymer Chelating Adsorbents

Polymer chelating adsorbents are developed to selectively adsorb heavy metals from aqueous systems. They are widely applied in preconcentration of trace elements from solutions for analytical purpose [52, 106–118]. Such adsorbents generally consist of the polymer matrix and the immobilized chelating groups to form complex with target heavy metals [119]. The specific interaction can be interpreted by Lewis acid–base principle, where the heavy metal ions can be taken as Lewis acids, while the chelating groups as Lewis bases. Heavy metals can be selectively adsorbed by chelate polymers, though strong affinity between the adsorbents and heavy metals makes it challenging and costly to regenerate the exhausted adsorbents for reuse. Besides, the chelating groups are normally unstable to oxidants, thereby losing the capability to chelate with heavy metals during long-term use.

Recent advancements on cation exchangers and chelating adsorbents employed for the removal of heavy metals are summarized in Table 5.6.

5.3.1.3 Polymer-Based Nanocomposite Adsorbents

In recent years, $M(HPO_4)_2$ (M = Zr, Ti, Sn) have been exploited as efficient adsorbents for heavy metal ions due to their cation exchange properties [129, 130]. However, $M(HPO_4)_2$ are usually present as fine or ultrafine particles for better adsorption performance. Unfortunately, the ultrafine particles cannot be readily used in fixed bed or any other flow-through systems due to the issues of being washed away and excessive pressure drop [131]. To overcome these technical issues, polymer nanocomposite adsorbents were developed. Recent researches [128, 132] indicated that polymer adsorbent or ion exchangers are promising hosts due to excellent mechanical strength and tunable surface chemistry. When a polymer cation exchanger is chosen to be the support material of amorphous zirconium phosphate (ZrP), the immobilized negatively charged functional group on the porous polymer matrix would greatly enhance permeation of heavy metal cation of counter charges because of Donnan membrane effect [105].

More specifically, ZrP was impregnated onto a strongly acidic cation exchange resin D001 and a nanocomposite adsorbent ZrP-001 was obtained [105]. Because the pore size of D-001 was mainly on nanoscale (the average pore size was ~34.1 nm), ZrP preloaded within D-001 was nanosized. ZrP-001 exhibited preferable adsorption toward lead ion over other nontoxic but ubiquitous cations like alkali metal or alkaline earth metal cation. In comparison with D-001, ZrP-001 exhibited more favorable lead adsorption even in the presence of Ca^{2+} or Mg^{2+} at much greater levels (Fig. 5.7). Such unique performance of ZrP-001 can be explained in two aspects, the improved diffusion kinetics caused by Donnan membrane effect [128] and the enhancement of adsorption selectivity caused by inner-sphere complex formation [129]. As for Donnan membrane effect, the immobilized negatively charged sulfonic group would greatly enhance the permeation and preconcentration of lead ion from aqueous phase to the polymer phase [128]. Moreover, FT-IR depicted in Fig. 5.8 indicates the newly formed Pb–O interaction after lead ion uptake, suggesting that the selective removal of lead mainly occurs through inner-sphere complex formation with ZrP [129]. In addition,

Table 5.6 Polymer cation exchangers and chelating adsorbents for the removal of heavy metals

Polymeric matrix	Functional group	Pollutant	Capacity	References
Polystyrene	Nitrosonaphthol	Cu, Ni	6–10 mmol/g	[120]
Polystyrene	Dithiooxamide	Cu	0.97 mmol/g	[121]
Polystyrene	Dithiooxamide	Zn, Pb	0.12 mmol/g	[121]
Polystyrene	Dithiooxamide	Cd	0.08 mmol/g	[121]
Polystyrene	Bis-2[(*O*-carbomethoxy)-phenoxy]ethylamine	La, Nd, Sm	0.5–0.7 mmol/g	[122]
Polystyrene	Palmitoyl quinolin-8-ol	Mn(II)	0.03 mmol/g	[123]
Polystyrene	1,2-bis(o-aminophenylthio)ethane	Hg	0.38 mmol/g	[124]
Polystyrene	1,2-bis(o-aminophenylthio)ethane	MeHg(I)	0.30 mmol/g	[124]
Polystyrene	Catechol	Cd, Cu, Ni, Pb	25–90 umol/g	[125]
Polystyrene	Amino group	Hg	0.8 mmol/g	[126]
		U	1.13 mmol/g	[126]
		Pb	0.36 mmol/g	[126]
Polystyrene	Bicine	La	0.35 mmol/g	[127]
		Nd	0.40 mmol/g	[127]
		Tb	0.42 mmol/g	[127]
		Th	0.25 mmol/g	[127]
Polystyrene	Sulfonic acid group	Pb(II)	323 mg/g	[128]
Iminodiacetate chelating resins	Iminodiacetate groups	Cu	3.26 mmol/g	[52]
		Co	2.60 mmol/g	[52]
		Ni	2.81 mmol/g	[52]
Poly(MMA-MAGA)	Methacryloylamidoglutamic acid groups	Hg	29.9 mg/g	[117]
		Cd	28.2 mg/g	[117]
		Pb	65.2 mg/g	[117]
Chelex 100	Iminodiacetic acid groups	Ni(II)	2.15 mmol/g	[118]

(continued)

Table 5.6 (continued)

Polymeric matrix	Functional group	Pollutant	Capacity	References
Poly(GMA-co-EGDMA)-en	Ethylene diamine	Pt(IV)	1.30 mmol/g	[106]
		Cu	1.10 mmol/g	[106]
		Pb	1.06 mmol/g	[106]
		Cd	0.67 mmol/g	[106]
GMA/DVB magnetic resin	Iminodiacetic acid groups	Pb	2.3 mmol/g	[112]
		Cd	2.0 mmol/g	[112]
		Zn	1.65 mmol/g	[112]
		Ca	1.60 mmol/g	[112]
		Mg	1.48 mmol/g	[112]
Dowex M 4195	Bispicolylamine groups	Cr(VI)	29.7 mg/g (Cr)	[114]
PASP chelating resin	Aspartate groups	Cu, Cd	1.3–1.40 mmol/g	[113]
Chelex 100	Iminodiacetic acid groups	Cu	1.6 mmol/g	[108]
Benzothiazole-based chelating resin	Benzothiazole groups	Cu	5.68 mmol/g	[110]
		Cd	1.03 mmol/g	[110]
		Pb	1.55 mmol/g	[110]
PGLY chelating resin	Glycine groups	Cu, Ni, Cd	~1.0 mmol/g	[109]
Azophenolcarboxylate-based chelating resin	Azophenolcarboxylate groups	Cr(III)	0.38 mmol/g	[111]
		Cr(VI)	0.69 mmol/g	[111]

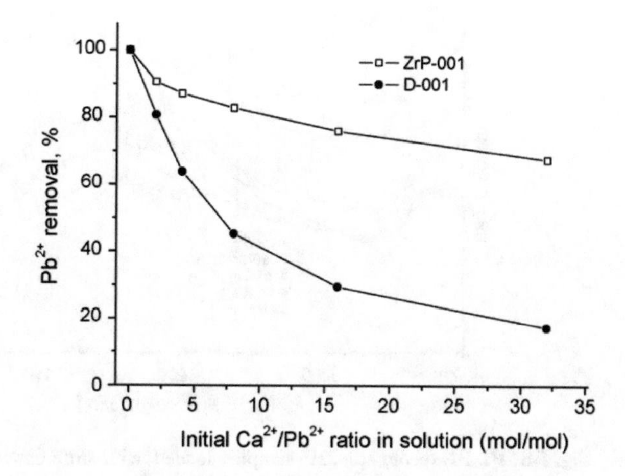

Fig. 5.7 Effect of Ca^{2+} on adsorption of Pb^{2+} onto ZrP-001 and D-001 [105]

efficient regeneration of exhausted ZrP-001 beads can be easily achieved by 2% HNO_3 or HCl solution at 303 K without any significant capacity loss.

Generally, inorganic NPs of specific interaction toward heavy metal ions, such as $M(HPO_4)_2$ (M = Zr, Ti), hydrated ferric oxides (HFOs), hydrous zirconium oxide (HZO), hydrated manganese oxide (HMO), $Zr(HPO_3S)_2$, can be loaded into porous polymer cation exchanger to obtain nanocomposite adsorbents specifically for heavy metals [133–135]. These materials are capable of incorporating high reactivity of NPs with easy operation of millimetric polystyrene beads, thus exhibiting great potential in advanced treatment of water contaminated by trace toxic metals.

5.3.2 *Polymer for Inorganic Anions*

Similar to heavy metal cations, inorganic anions can also be removed from aqueous system by ion exchange. Generally, anion exchangers contain positively charged groups such as primary amine, tertiary amine, or quaternary ammonium groups as functional groups. As nonselective electrostatic attraction was mainly involved, adsorption of target anionic pollutants by traditional ion exchangers would be suppressed intensively by massive coexisting anions.

Many environmentally benign inorganic NPs, such as metal (hydr)oxides, can preferably sequestrate various anionic pollutants by forming inner-sphere complex. In the past decade, millimetric polymers have become one of the most promising hosts to support inorganic NPs in terms of excellent mechanical strength, fine hydraulic property, and adjustable surface chemistry. Similarly, positively charged functional groups could preconcentrate anionic pollutants due to Donnan effect. Preferable adsorption of target pollutants mainly resulted from the embedded NPs, and some examples using polymer nanocomposites to remove anionic pollutants from water are summarized in Table 5.7.

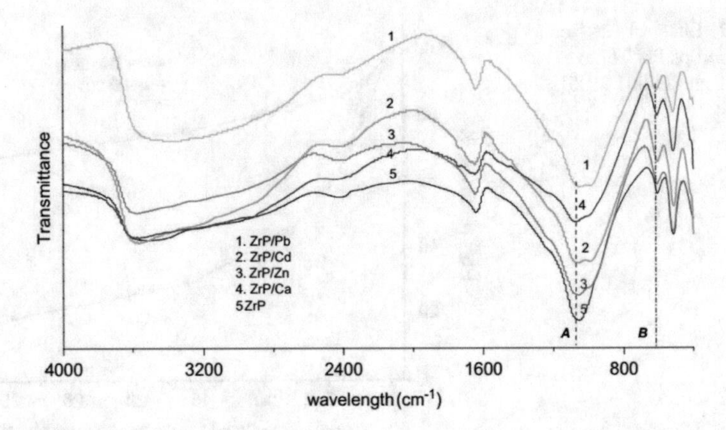

Fig. 5.8 FT-IR spectra of ZrP samples loaded with different metal ions obtained at solution equilibrium pH of 4.4–4.7 and 303 K. [129]

The encapsulated NPs often exhibit distinct properties from the bulky ones. For instance, hydrated manganese oxide (HMO) is among the most abundant minerals in the earth's crust, though it is usually negatively charged at circumneutral pHs and thus represents an unfavorable adsorbent for anionic pollutants like phosphate. Pan et al. [142] proposed a new strategy to sequestrate phosphate by using nano-HMO encapsulated inside a polystyrene anion exchanger (NS). As shown in Fig. 5.9, HMO NPs are successfully loaded and uniformly dispersed inside the NS beads with the average particle size of 5.0–7.0 nm. In this way, the pH_{pzc} shifted from 6.2 for the bulky HMO to 10.5 for the capsulated HMO nanoparticles, leaving HMO NPs positively charged at neutral pH. HMO@NS sample was thus capable of generating 460 bed volume (BV) clean water containing phosphate <2 mg/L by forming reversible and irreversible complex, while the polymer host NS could only generate $\sim$80 BV clean water under otherwise identical conditions. The nanocomposite adsorbent can be regenerated by NaOH–NaCl solution. Another example is about the regeneration of the exhausted lanthanum (hydr)oxide (HLO). HLO could provide great amount of coordination sites and specific affinity for phosphate even at trace levels. However, it remains a great challenge to refresh phosphate-loaded HLO under mild conditions arising from the extremely strong interaction between HLO and phosphate. Zhang et al. [99] immobilized hydrated La (III) oxide (HLO) nanoclusters inside a polystyrene anion exchanger D-201 to form a new nanocomposite La-201. La-201 exhibited amazing adsorption toward phosphate. For instance, it can treat $\sim$6500 BV phosphate-polluted water, approximately 11 times higher magnitude than that of HFO-201, a commercial nanocomposite with similar structure, except that HFO NPs were encapsulated inside polymer matrix. The exhausted La-201 could be regenerated with NaOH–NaCl binary solution at 60 °C for repeated use without any significant capacity loss. The underlying mechanism for the specific sorption of phosphate by La-201 was revealed with the aid of STEM-EDS, XPS, XRD, and SSNMR analysis, and cyclic

Table 5.7 Polymer nanocomposite adsorbent for the removal of anionic pollutants

Polymeric matrix	NPs	Pollutant	Adsorption performance	References
Polymer anion exchangers	Hydrated ferric oxide	Arsenic	As(V) removed from 23 to <0.5 ppb within 33,196 BV	[136]
Polymer anion exchangers	Hydrated ferric oxide	Arsenic	As(V) removed from 50 to <10 ppb within 4000 BV, As(III) removed from 100 to <10 ppb within 2000 BV	[137]
Polymer anion exchangers	Hydrated ferric oxide	Arsenic	As(V) removed from 300 to <10 ppb within 3500 BV, As(V) removed from 20 to <10 ppb within 17,500 BV	[138]
Polystyrene adsorbents	Hydrated ferric oxide	Arsenic	Arsenic removed from 100 to <10 ppb within 60 BV	[139]
Polymer anion exchangers D-201	Hydrated zirconium oxide	Arsenic	Arsenate removed from 100 to <10 ppb within 2600 BV	[140]
Polymer anion exchangers D-201	Aged zero-valent iron	Arsenic	As(III) removed by 80% and As(V) removed by 100% after 400 min	[141]
Polymer anion exchangers	Hydrated ferric oxide	Phosphate	P(V) removed from 100 to <5 ppb within 10,000 BV	[132]
Polystyrene anion exchanger	Hydrous manganese oxide	Phosphate	P(V) removed from 2 to 0.5 mg P[PO_4^{3-}]/L within ~460 BV	[142]
Polymer anion exchangers D-201	Hydrated lanthanum oxide	Phosphate	P(V) removed from 2.5 to <0.5 mg P/L within ~6500 BV	[99]
Polymer anion exchangers D-201	Hydrated zirconium oxide	Fluoride	F removed from around 3.5 to <1 mg F−/L within 3000 BV	[95]
Active chlorine covalently binding spherical polystyrene adsorbents	Hydrated ferric oxide	Arsenic	As(III) removed from 200 to <10 μg/L within ~1760 BV	[101]
Hyper-cross-linked polystyrene anion exchanger binding tertiary amine groups	Hydrous zirconium oxide	Fluoride	F removed from 3.3 to <1.5 mg F/L within ~80 BV	[100]

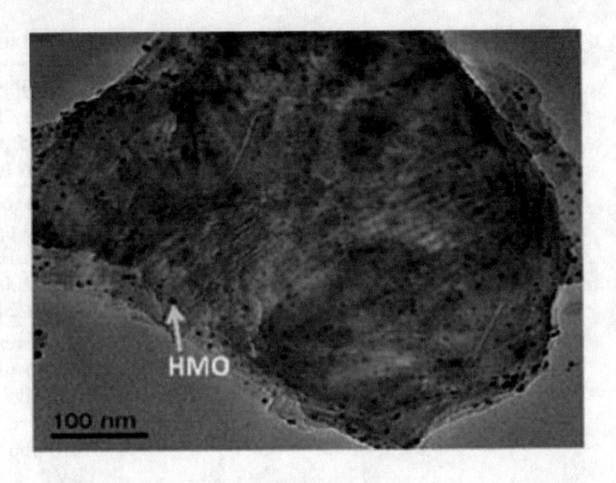

Fig. 5.9 TEM image of HMO@NS nanocomposite adsorbent [142]

phase transition from $LaPO_4 \times H_2O$ to HLO is verified to be the dominant pathway for outstanding regeneration property of La-201.

Rational design of antifouling polymer nanocomposite represents an attractive strategy to further improve their applicability in practical water treatment. Zhang et al. [100] developed a novel nanocomposite adsorbent HZO@HCA by encapsulating nanosized hydrous zirconium oxide (HZO) inside hyper-cross-linked polystyrene binding tertiary amine groups. Since the adsorbent features with abundant micropores instead of meso-/macropores, natural organic matters (NOM) of large size are incapable of diffusing inside the nanocomposite adsorbent due to size exclusion, thereby avoiding undesirable interactions with the embedded NPs. Moreover, tertiary amine groups facilitate desorption of NOM from HZO@HCA, because they turned to be negatively charged under alkaline pHs and thus repulsed negatively charged NOM molecules.

5.4 Summary

Polymer adsorbents and their derivatives have been developed for over seven decades and used widely in polluted water treatment, such as organic wastewater treatment and advanced removal of heavy metals. However, there are still many challenging issues on the synthesis and application of polymer adsorbents in environmental application. To be specific, although supporting matrix, pore structure, and functional groups of polymer adsorbents could be tailor-made, molecular design of a polymer adsorbent for highly specific adsorption toward a given pollutant is still difficult. Also, recovery of the concentrated eluate is still a costly task. In addition, adsorption capacities of polymer adsorbents toward soluble pollutants, i.e., ionizable organic compounds, are relatively low, and frequent regeneration is thus required, which will greatly bring up the operation cost in massive application. As for the polymer nanocomposite adsorbents, the precise manipulation on their structure, including pore structure, NPs size and distribution, and functional group is very attractive but still challenging. Thus, the improvement of working capacity of polymer and polymer nanocomposite adsorbents, especially the adsorption capacity, adsorption dynamics, and selectivity, remains the key and core for further study [102]. Besides, the adsorption mechanism toward target species needs to be elucidated not only from the overall experimental evidences or crude interaction force summarization, but also from molecular-level interactions. Also, mathematical modeling is required to optimize the fixed-bed operation of polymer adsorbents in scaled-up application.

References

1. Jang M, Chen W, Cannon FS (2008) Preloading hydrous ferric oxide into granular activated carbon for arsenic removal. Environ Sci Technol 42(9):3369–3374
2. Vaughan RL, Reed BE (2005) Modeling As(V) removal by a iron oxide impregnated activated carbon using the surface complexation approach. Water Res 39(6):1005–1014
3. Zhuang JM, Hobenshield E, Walsh T (2008) Arsenate sorption by hydrous ferric oxide incorporated onto granular activated carbon with phenol formaldehyde resins coating. Environ Technol 29(4):401–411
4. Guo XJ, Chen FH (2005) Removal of arsenic by bead cellulose loaded with iron oxyhydroxide from groundwater. Environ Sci Technol 39(17):6808–6818
5. Chen KL, Mylon SE, Elimelech M (2007) Enhanced aggregation of alginate-coated iron oxide (hematite) nanoparticles in the presence of calcium, strontium, and barium cations. Langmuir 23(11):5920–5928
6. Zouboulis AI, Katsoyiannis IA (2002) Arsenic removal using iron oxide loaded alginate beads. Ind Eng Chem Res 41(24):6149–6155
7. Jang M, Min S-H, Park JK, Tlachac EJ (2007) Hydrous ferric oxide incorporated diatomite for remediation of arsenic contaminated groundwater. Environ Sci Technol 41(9):3322–3328
8. Jang M, Min SH, Kim TH, Park JK (2006) Removal of arsenite and arsenate using hydrous ferric oxide incorporated into naturally occurring porous diatomite. Environ Sci Technol 40 (5):1636–1643
9. Kalderis D, Koutoulakis D, Paraskeva P, Diamadopoulos E, Otal E, Olivares del Valle J, Fernandez-Pereira C (2008) Adsorption of polluting substances on activated carbons prepared from rice husk and sugarcane bagasse. Chem Eng J 144(1):42–50
10. Qiu Y, Cheng H, Xu C, Sheng D (2008) Surface characteristics of crop-residue-derived black carbon and lead(II) adsorption. Water Res 42(3):567–574
11. Yu Z, Peldszus S, Huck PM (2008) Adsorption characteristics of selected pharmaceuticals and an endocrine disrupting compound—naproxen, carbamazepine and nonylphenol—on activated carbon. Water Res 42(12):2873–2882
12. Karaca S, Gürses A, Açikyildiz M, Ejder M (2008) Adsorption of cationic dye from aqueous solutions by activated carbon. Microporous Mesoporous Mater 115(3):376–382
13. Karanfil T, Dastgheib SA (2004) Trichloroethylene adsorption by fibrous and granular activated carbons: aqueous phase, gas phase, and water vapor adsorption studies. Environ Sci Technol 38(22):5834–5841
14. Hernandez-Ramirez O, Holmes SM (2008) Novel and modified materials for wastewater treatment applications. J Mater Chem 18(24):2751–2761
15. Memon SQ, Memon N, Solangi AR, Memon J-u-R (2008) Sawdust: a green and economical sorbent for thallium removal. Chem Eng J 140(1–3):235–240

16. Ríos CA, Williams CD, Roberts CL (2008) Removal of heavy metals from acid mine drainage (AMD) using coal fly ash, natural clinker and synthetic zeolites. J Hazard Mater 156(1–3):23–35
17. Tripathy SS, Raichur AM (2008) Abatement of fluoride from water using manganese dioxide-coated activated alumina. J Hazard Mater 153(3):1043–1051
18. Vilar VJP, Botelho CMS, Boaventura RAR (2008) Metal biosorption by algae Gelidium derived materials from binary solutions in a continuous stirred adsorber. Chem Eng J 141(1–3):42–50
19. Pan B, Pan B, Zhang W, Lv L, Zhang Q, Zheng S (2009) Development of polymeric and polymer-based hybrid adsorbents for pollutants removal from waters. Chem Eng J 151(1–3):19–29
20. Pan BC, Zhang QX, Meng FW, Li XT, Zhang X, Zheng JZ, Zhang WM, Pan BJ, Chen JL (2005) Sorption enhancement of aromatic sulfonates onto an aminated hyper-cross-linked polymer. Environ Sci Technol 39(9):3308–3313
21. Sun Y, Chen J, Li A, Liu F, Zhang Q (2005) Adsorption of resorcinol and catechol from aqueous solution by aminated hypercrosslinked polymers. React Funct Polym 64(2):63–73
22. Bratkowska D, Fontanals N, Borrull F, Cormack PA, Sherrington DC, Marce RM (2010) Hydrophilic hypercrosslinked polymeric sorbents for the solid-phase extraction of polar contaminants from water. J Chromatogr A 1217(19):3238–3243
23. Garcia A, Ferreira L, Leitao A, Rodrigues A (1999) Binary adsorption of phenol and m-cresol mixtures onto a polymeric adsorbent. Adsorpt. J. Int. Adsorpt. Soc. 5(4):359–368
24. He C, Huang K, Huang J (2010) Surface modification on a hyper-cross-linked polymeric adsorbent by multiple phenolic hydroxyl groups to be used as a specific adsorbent for adsorptive removal of p-nitroaniline from aqueous solution. J Colloid Interface Sci 342 (2):462–466
25. Bilgili MS (2006) Adsorption of 4-chlorophenol from aqueous solutions by XAD-4 resin: isotherm, kinetic, and thermodynamic analysis. J Hazard Mater 137(1):157–164
26. Xu ZY, Zhang QX, Wu CL, Wang LS (1997) Adsorption of naphthalene derivatives on different macroporous polymeric adsorbents. Chemosphere 35(10):2269–2276
27. Long C, Li A, Wu H, Liu F, Zhang Q (2008) Polanyi-based models for the adsorption of naphthalene from aqueous solutions onto nonpolar polymeric adsorbents. J Colloid Interface Sci 319(1):12–18
28. Lee JW, Jun HJ, Kwak DH, Chung PG (2005) Adsorption of dichloromethane from water onto a hydrophobic polymer resin XAD-1600. Water Res 39(4):617–629
29. Zheng K, Pan B, Zhang Q, Zhang W, Pan B, Han Y, Zhang Q, Wei D, Cu Z, Zhang Q (2007) Enhanced adsorption of p-nitroaniline from water by a carboxylated polymeric adsorbent. Sep Purif Technol 57(2):250–256
30. Otero M, Zabkova M, Rodrigues AE (2005) Comparative study of the adsorption of phenol and salicylic acid from aqueous solution onto nonionic polymeric resins. Sep Purif Technol 45(2):86–95
31. Yang WC, Shim WG, Lee JW, Moon H (2003) Adsorption and desorption dynamics of amino acids in a nonionic polymeric sorbent XAD-16 column. Korean J Chem Eng 20 (5):922–929
32. Deosarkar SP, Pangarkar VG (2004) Adsorptive separation and recovery of organics from PHBA and SA plant effluents. Sep Purif Technol 38(3):241–254
33. Abburi K (2003) Adsorption of phenol and p-chlorophenol from their single and bisolute aqueous solutions on Amberlite XAD-16 resin. J Hazard Mater 105(1–3):143–156
34. Musty PR, Nickless G (1974) Use of amberlite XAD-4 for extraction and recovery of chlorinated insecticides and polychlorinated biphenyls from water. J Chromatogr 89(2): 185–190
35. Hua M, Zhang S, Pan B, Zhang W, Lv L, Zhang Q (2012) Heavy metal removal from water/wastewater by nanosized metal oxides: a review. J Hazard Mater 211–212(Suppl. C):317–331

36. Zhao X, Lv L, Pan B, Zhang W, Zhang S, Zhang Q (2011) Polymer-supported nanocomposites for environmental application: a review. Chem Eng J 170(2–3):381–394
37. Swallow KC, Hume DN, Morel FMM (1980) Sorption of copper and lead by hydrous ferric-oxide. Environ Sci Technol 14(11):1326–1331
38. Kinniburgh DG, Jackson ML, Syers JK (1976) Adsorption of alkaline-earth, transition, and heavy-metal cations by hydrous oxide gels of iron and aluminum. Soil Sci Soc Am J 40 (5):796–799
39. Fan M, Boonfueng T, Xu Y, Axe L, Tyson TA (2005) Modeling Pb sorption to microporous amorphous oxides as discrete particles and coatings. J Colloid Interface Sci 281(1):39–48
40. Trivedi P, Axe L, Tyson TA (2001) XAS studies of Ni and Zn sorbed to hydrous manganese oxide. Environ Sci Technol 35(22):4515–4521
41. Bargar JR, Brown GE, Parks GA (1997) Surface complexation of Pb(II) at oxide-water interfaces. 1. XAFS and bond-valence determination of mononuclear and polynuclear Pb(II) sorption products on aluminum oxides. Geochim Cosmochim Acta 61(13):2617–2637
42. Jang J-H, Dempsey BA (2008) Coadsorption of arsenic (III) and arsenic (V) onto hydrous ferric oxide: effects on abiotic oxidation of arsenic (III), extraction efficiency, and model accuracy. Environ Sci Technol 42(8):2893–2898
43. Kawashima M, Tainaka Y, Hori T, Koyama M, Takamatsu T (1986) Phosphate adsorption onto hydrous manganese(IV) oxide in the presence of divalent-cations. Water Res 20 (4):471–475
44. Streat M, Sweetland LA (1998) Removal of pesticides from water using hypercrosslinked polymer phases. Process Saf Environ Prot 76(2):127–134
45. Zhaoyi X, Quauxing Z, Changlong W, Liansheng W (1997) Adsorption of naphthalene derivatives on different macroporous polymeric adsorbents. Chemosphere 38(10):8
46. Yang WC, Shim WG, Moon JWLAH (2003) Adsorption and desorption dynamics of amino acids in a nonionic polymeric sorbent XAD-16 column. Korean J Chem Eng 20(5):8
47. Pan B, Du W, Zhang W, Zhang X, Zhang Q, Pan B, Lv L, Zhang Q, Chen J (2007) Improved adsorption of 4-nitrophenol onto a novel hyper-cross-linked polymer. Environ Sci Technol 41(14):5057–5062
48. Pan B, Zhang W, Pan B, Qiu H, Zhang Q, Zhang Q, Zheng S (2008) Efficient removal of aromatic sulfonates from wastewater by a recyclable polymer: 2-naphthalene sulfonate as a representative pollutant. Environ Sci Technol 42(19):7411–7416
49. Masque N, Galia M, Marce RM, Borrull F (1999) Influence of chemical modification of polymeric resin on retention of polar compounds in solid-phase extraction. Chromatographia 50(1–2):21–26
50. Xu Z, Zhang Q, Fang HHP (2003) Applications of porous resin sorbents in industrial wastewater treatment and resource recovery. Crit Rev Environ Sci Technol 33(4):363–389
51. Kunin R (1980) Porous polymers as adsorbents—a review of current practice. Anzber-lzi-lites, p 163
52. Dinu MV, Dragan ES (2008) Heavy metals adsorption on some iminodiacetate chelating resins as a function of the adsorption parameters. React Funct Polym 68(9):1346–1354
53. Pan B, Zhang W, Zhang Q, Zheng S (2008) Adsorptive removal of phenol from aqueous phase by using a porous acrylic ester polymer. J Hazard Mater 157(2–3):293–299
54. Okay O (2000) Macroporous copolymer networks. Prog Polym Sci (Oxford) 25(6):711–779
55. Simpson EJ, Abukhadra RK, Koros WJ, Schechter RS (1993) Sorption equilibrium isotherms for volatile organics in aqueous solution: comparison of head-space gas chromatography and on-line UV stirred cell results. Ind Eng Chem Res 32(10):2269–2276
56. Long C, Li A, Gao G, Fei Z, Zhang Q, Chen J, Reclaiming technique by using resin to adsorb nitro chlorobenzene in wastewater from producing nitro chlorobenzene. CN1562789-A, CN1233570-C
57. Freitas PA, Iha K, Felinto MC, Suarez-Iha ME (2008) Adsorption of di-2-pyridyl ketone salicyloylhydrazone on amberlite XAD-2 and XAD-7 resins: characteristics and isotherms. J Colloid Interface Sci 323(1):1–5

58. Zhang W, Xu Z, Pan B, Hong C, Jia K, Jiang P, Zhang Q (2008) Equilibrium and heat of adsorption of diethyl phthalate on heterogeneous adsorbents. J Colloid Interface Sci 325 (1):41–47
59. Wang G, Dou BJ, Wang JH, Wang WQ, Hao ZP (2013) Adsorption properties of benzene and water vapor on hyper-cross-linked polymers. RSC Adv 3(43):20523–20531
60. Zhang W, Xu Z, Zhang Q, Pan B, Du W, Chen J, Method for treating diethyl (o-) phthalate waste water and recovering diethyl (o-) phthalate from it. CN1935776-A, CN100453523-C
61. Davankov VA, Rogoshin SV, Tsyurupa MP (1974) Macronet isoporous gels through crosslinking of dissolved polystyrene. J Polym Sci Part C Polym Symp 47:95–101
62. Oh CG, Ahn JH, Ihm SK (2003) Adsorptive removal of phenolic compounds by using hypercrosslinked polystyrenic beads with bimodal pore size distribution. React Funct Polym 57(2–3):103–111
63. Li A, Zhang Q, Zhang G, Chen J, Fei Z, Liu F (2002) Adsorption of phenolic compounds from aqueous solutions by a water-compatible hypercrosslinked polymeric adsorbent. Chemosphere 47(9):981–989
64. Nastaj J, Kamińska A (2008) Adsorption of phenol on water-fluidized polymeric amberlite XAD-4 and XAD-16 adsorbents. Przem Chem 87(3):300–306
65. Azanova VV, Hradil J (1999) Sorption properties of macroporous and hypercrosslinked copolymers. React Funct Polym 41(1):163–175
66. Juang RS, Shiau JY (1999) Adsorption isotherms of phenols from water onto macroreticular resins. J Hazard Mater 70(3):171–183
67. Gusler GM, Browne TE, Cohen Y (1993) Sorption of organics from aqueous solution onto polymeric resins. Ind Eng Chem Res 32(11):2727–2735
68. Yang W, Li A, Fu C, Fan J, Zhang Q (2007) Adsorption mechanism of aromatic sulfonates onto resins with different matrices. Ind Eng Chem Res 46(21):6971–6977
69. Saikia MD (2008) Revisiting adsorption of biomolecules on polymeric resins. Colloids Surf A 315(1–3):196–204
70. Bohra PM, Vaze AS, Pangarkar VG, Taskar A (1994) Adsorptive recovery of water-soluble essential oil components. J Chem Technol Biotechnol 60(1):97–102
71. Streat M, Sweetland LA (1997) Physical and adsorptive properties of hypersol-macronet™ polymers. React Funct Polym 35(1–2):99–109
72. Davankov VA, Rogozhin SV, Tsyurupa MP (1969) Macronet polystyren strucutres for ionites and method of producing same. US Patent, 3729457
73. Davankov VA, Rogozhin SV, Tsyurupa MP (1973) Macronet polystyrene structures for ionites and method of producing same. In: Macronet polystyrene structures for ionites and method of producing same
74. Zeng X, Fan Y, Wu G, Wang C, Shi R (2009) Enhanced adsorption of phenol from water by a novel polar post-crosslinked polymeric adsorbent. J Hazard Mater 169(1–3):1022–1028
75. Bai LL, Zhou YH, Wang XL, Yuan SG, Wu XL (2011) Facile synthesis of hypercrosslinked resin via photochlorination of p-xylene and succedent alkylation polymerization. Chin Chem Lett 22(9):1115–1118
76. Xiaohui Z, Siguo Y (2011) Adsorption of benzene from air, solution and film floating on the water by non-polystyrenr hypercrosslinked resin. Ion Exch Adsorpt 27(4):297–303
77. Jafvert CT, Westall JC, Grieder E, Schwarzenbach RP (1990) Distribution of hydrophobic ionogenic organic compounds between octanol and water: organic acids. Environ Sci Technol 24(12):1795–1803
78. Stapleton MG, Sparks DL, Dentel SK (1994) Sorption of pentachlorophenol to HDTMA-clay as a function of ionic strength and pH. Environ Sci Technol 28(13):2330–2335
79. Zhang WM, Pan BC, Xu ZW, Hong CH, Zhang QJ, Zhang B, Li AM, Pan BG, Zhang QX, Chen JL (2007) Method of increasing hydrophilicity of complex function adsorption resin and reinforcing adsorbability of the complex function adsorption resin. Chinese Patent, CN 20071001997.3

80. Xiao G, Fu L, Li A (2012) Enhanced adsorption of bisphenol A from water by acetylaniline modified hyper-cross-linked polymeric adsorbent: effect of the cross-linked bridge. Chem Eng J 191:171–176
81. Pan BC, Xiong Y, Li AM, Chen JL, Zhang QX, Jin XY (2002) Adsorption of aromatic acids on an aminated hypercrosslinked macroporous polymer. React Funct Polym 53(2–3):63–72
82. Wang RF, Shi ZQ, Shi RF, Zhang JZ, Ou LL (2005) The study of adsorption of phenol and aniline on aminated-macroporous hypercrosslinked resins. Acta Polym Sin 3:339–344
83. Chang C-F, Chang C-Y, Hsu K-E, Lee S-C, Hoell W (2008) Adsorptive removal of the pesticide methomyl using hypercrosslinked polymers. J Hazard Mater 155(1–2):295–304
84. Yu Y, Zhuang YY, Wang ZH (2001) Adsorption of water-soluble dye onto functionalized resin. J Colloid Interface Sci 242(2):288–293
85. Li AM, Cai JG, Zhang HY, Ge JJ, Li ZB, Long C, Liu FQ, Zhang QX (2005) Chinese patent: CN 1712365A
86. Chen JL, Pan BC, Xiong Y, Li AM, Long C, Han YZ, Sun Y, Zhang QX (2002) Chinese patent: CN1384069
87. Lee YS, Ryoo SJ (2002) US patent: US6369169
88. Zhu XX, Brizard F, Piche J, Yim CT, Brown GR (2000) Bile salt anion sorption by polymeric resins: comparison of a functionalized polyacrylamide resin with cholestyramine. J Colloid Interface Sci 232(2):282–288
89. Zhang Q, Li A, Liu F (2004) Synthesizing weakly alkaline anionic exchange resin with double functions and superhigh cross-linking. CN1346708-A, CN1131112-C
90. Pan B, Zhang Q, Pan B, Zhang W, Du W, Ren H (2008) Removal of aromatic sulfonates from aqueous media by aminated polymeric sorbents: concentration-dependent selectivity and the application. Microporous Mesoporous Mater 116(1–3):63–69
91. Chen J, Pan B, Xiong Y, Method of treating naphthalene-blowing effluence and recovering resource in 2-naphthol producing process. CN1384069-A, CN1139539-C
92. Zhang Q, Long C, Xu Z, Long CXZ, Treatment and rediaimation of waste water in production of 4,4′-dinitrobistyrene-2,2′-bisulfonic acid. CN1304882-A, CN1156407-C
93. Hua M, Zhang S, Pan B, Zhang W, Lv L, Zhang Q (2012) Heavy metal removal from water/ wastewater by nanosized metal oxides: a review. J Hazard Mater 211:317–331
94. Jiang Y, Hua M, Wu B, Ma H, Pan B, Zhang Q (2014) Enhanced removal of arsenic from a highly laden industrial effluent using a combined coprecipitation/nano-adsorption process. Environ Sci Pollut Res 21(10):6729–6735
95. Pan B, Xu J, Wu B, Li Z, Liu X (2013) Enhanced removal of fluoride by polystyrene anion exchanger supported hydrous zirconium oxide nanoparticles. Environ Sci Technol 47 (16):9347–9354
96. Jiang Z, Zhang S, Pan B, Wang W, Wang X, Lv L, Zhang W, Zhang Q (2012) A fabrication strategy for nanosized zero valent iron (nZVI)-polymeric anion exchanger composites with tunable structure for nitrate reduction. J Hazard Mater 233:1–6
97. Zhang Q, Du Q, Jiao T, Pan B, Zhang Z, Sun Q, Wang S, Wang T, Gao F (2013) Selective removal of phosphate in waters using a novel of cation adsorbent: Zirconium phosphate (ZrP) behavior and mechanism. Chem Eng J 221:315–321
98. Xie Y, Lv L, Zhang S, Pan B, Wang X, Chen Q, Zhang W, Zhang Q (2011) Fabrication of anion exchanger resin/nano-CdS composite photocatalyst for visible light RhB degradation. Nanotechnology 22(30):305707
99. Zhang Y, Pan B, Shan C, Gao X (2016) Enhanced phosphate removal by nanosized hydrated La(III) oxide confined in cross-linked polystyrene networks. Environ Sci Technol 50 (3):1447–1454
100. Zhang X, Zhang L, Li Z, Jiang Z, Zheng Q, Lin B, Pan B (2017) Rational design of antifouling polymeric nanocomposite for sustainable fluoride removal from NOM-rich water. Environ Sci Technol 51(22):13363–13371
101. Zhang X, Wu M, Dong H, Li H, Pan B (2017) Simultaneous oxidation and sequestration of As(III) from water by using redox polymer-based Fe(III) oxide nanocomposite. Environ Sci Technol 51(11):6326–6334

102. Zhang Q, Li A, Pan B (2015) The development of ion exchange and adsorption resin and its application in industrial wastewater treatment and resource reuse. Polym Bull 9:21–43
103. Goto M, Goto S (1987) Removal and recovery of heavy-metals by ion-exchange fiber. J Chem Eng Jpn 20(5):467–472
104. Egen N, Ford PC, Grotz LC (1976) Hard water, water softening, ion-exchange. J Chem Educ 53(5):302–303
105. Zhang QR, Du W, Pan BC, Pan BJ, Zhang WM, Zhang QJ, Xu ZW, Zhang QX (2008) A comparative study on Pb^{2+}, Zn^{2+} and Cd^{2+} sorption onto zirconium phosphate supported by a cation exchanger. J Hazard Mater 152(2):469–475
106. Nastasović A, Jovanović S, Đorđević D, Onjia A, Jakovljević D, Novaković T (2004) Metal sorption on macroporous poly(GMA-co-EGDMA) modified with ethylene diamine. React Funct Polym 58(2):139–147
107. Hosseini MS, Raissi H, Madarshahian S (2006) Synthesis and application of a new chelating resin functionalized with 2,3-dihydroxy benzoic acid for Fe(III) determination in water samples by flame atomic absorption spectrometry. React Funct Polym 66(12):1539–1545
108. Alberti G, Pesavento M, Biesuz R (2007) A chelating resin as a probe for the copper(II) distribution in grape wines. React Funct Polym 67(10):1083–1093
109. Chen C, Chiang C (2007) Removal of heavy metal ions by a chelating resin containing glycine as chelating groups. Sep Purif Technol 54(3):396–403
110. Meesri S, Praphairaksit N, Imyim A (2007) Extraction and preconcentration of toxic metal ions from aqueous solution using benzothiazole-based chelating resins. Microchem J 87 (1):47–55
111. Pramanik S, Dey S, Chattopadhyay P (2007) A new chelating resin containing azophenol-carboxylate functionality: synthesis, characterization and application to chromium speciation in wastewater. Anal Chim Acta 584(2):469–476
112. Atia AA, Donia AM, Yousif AM (2008) Removal of some hazardous heavy metals from aqueous solution using magnetic chelating resin with iminodiacetate functionality. Sep Purif Technol 61(3):348–357
113. Chen CY, Lin MS, Hsu KR (2008) Recovery of Cu(II) and Cd(II) by a chelating resin containing aspartate groups. J Hazard Mater 152(3):986–993
114. Saygi KO, Tuzen M, Soylak M, Elci L (2008) Chromium speciation by solid phase extraction on Dowex M 4195 chelating resin and determination by atomic absorption spectrometry. J Hazard Mater 153(3):1009–1014
115. Burke WA, Removal of heavy metal cation contaminants from organic soln.-using chelating ion exchange resin modified by removal of sodium ions. US5525315-A
116. Schneider HP, Wallbaum U (1990) Gel-type chelating resins and a process for removal of multi-valent, alkaline earth or heavy metal cations from solutions 4895905
117. Denizli A, Sanli N, Garipcan B, Patir S, Alsancak G (2004) Methacryloylamidoglutamic acid incorporated porous poly(methyl methacrylate) beads for heavy-metal removal. Ind Eng Chem Res 43(19):6095–6101
118. Leinonen H, Lehto J (2000) Ion-exchange of nickel by iminodiacetic acid chelating resin Chelex 100. React Funct Polym 43(1–2):1–6
119. Dabrowski A, Hubicki Z, Podkoscielny P, Robens E (2004) Selective removal of the heavy metal ions from waters and industrial wastewaters by ion-exchange method. Chemosphere 56(2):91–106
120. Memon SQ, Bhanger MI, Hasany SM, Khuhawar MY (2007) The efficacy of nitrosonaphthol functionalized XAD-16 resin for the preconcentration/sorption of Ni(II) and Cu(II) ions. Talanta 72(5):1738–1745
121. Dutta S, Das AK (2007) Synthesis, characterization, and application of a new chelating resin functionalized with dithiooxamide. J Appl Polym Sci 103(4):2281–2285
122. Kaur H, Agrawal YK (2005) Functionalization of XAD-4 resin for the separation of lanthanides using chelation ion exchange liquid chromatography. React Funct Polym 65 (3):277–283

123. Dogutan M, Filik H, Apak R (2003) Preconcentration of manganese(II) from natural and sea water on a palmitoyl quinolin-8-ol functionalized XAD copolymer resin and spectrophotometric determination with the formaldoxime reagent. Anal Chim Acta 485(2):205–212
124. Mondal BC, Das AK (2003) Determination of mercury species with a resin functionalized with a 1,2-bis(o-aminophenylthio)ethane moiety. Anal Chim Acta 477(1):73–80
125. Bernard J, Branger C, Nguyen TLA, Denoyel R, Margaillan A (2008) Synthesis and characterization of a polystyrenic resin functionalized by catechol: application to retention of metal ions. React Funct Polym 68(9):1362–1370
126. Rivas BL, Pooley SA, Maturana HA, Villegas S (2001) Sorption properties of poly (styrene-co-divinylbenzene) amine functionalized weak resin. J Appl Polym Sci 80 (12):2123–2127
127. Dev K, Pathak R, Rao GN (1999) Sorption behaviour of lanthanum(III), neodymium(III), terbium(III), thorium(IV) and uranium(VI) on Amberlite XAD-4 resin functionalized with bicine ligands. Talanta 48(3):579–584
128. Pan BC, Zhang QR, Zhang WM, Pan BJ, Du W, Lv L, Zhang QJ, Xu ZW, Zhang QX (2007) Highly effective removal of heavy metals by polymer-based zirconium phosphate: a case study of lead ion. J Colloid Interface Sci 310(1):99–105
129. Pan B, Zhang Q, Du W, Zhang W, Xu Z (2007) Selective heavy metals removal from waters by amorphous zirconium phosphate: behavior and mechanism. Water Res 41(14):3103–3111
130. Jia K, Pan B, Zhang Q, Zhang W, Jiang P, Hong C (2008) Adsorption of Pb^{2+}, Zn^{2+}, and Cd^{2+} from waters by amorphous titanium phosphate. J Colloid Interface Sci 318(2):160–166
131. Cumbal L, Sengupta AK (2005) Arsenic removal using polymer-supported hydrated iron (III) oxide nanoparticles: Role of Donnan membrane effect. Environ Sci Technol 39 (17):6508–6515
132. Blaney LM, Cinar S, SenGupta AK (2007) Hybrid anion exchanger for trace phosphate removal from water and wastewater. Water Res 41(7):1603–1613
133. Zhang Q, Pan B, Chen X, Zhang W, Lv L, Zhao XS (2008) Preparation of polymer-supported hydrated ferric oxide based on Donnan membrane effect and its application for arsenic removal. Sci China Ser B Chem 51(4):379–385
134. Zhang Q, Pan B, Zhang W, Jia K (2008) Selective sorption of lead, cadmium and zinc ions by a polymeric cation exchanger containing nano-$Zr(HPO_3S)_2$. Environ Sci Technol 42 (11):4140–4145
135. Pan BC, Su Q, Zhang WM, Zhang QX, Ren HQ, Zhang QJ, Zhang QR, Pan BJ (2007) Enviromental functional material based on nanoparticles hydrated manganese oxide and preparing method thereof. Chinese Patent, CN20071013405.0
136. Sylvester P, Westerhoff P, Moller T, Badruzzaman M, Boyd O (2007) A hybrid sorbent utilizing nanoparticles of hydrous iron oxide for arsenic removal from drinking water. Environ Eng Sci 24(1):104–112
137. DeMarco MJ, Sengupta AK, Greenleaf JE (2003) Arsenic removal using a polymeric/inorganic hybrid sorbent. Water Res 37(1):164–176
138. Moller T, Sylvester P (2008) Effect of sililca and pH on arsenic uptake by resin/iron oxide hybrid media. Water Res 42(6–7):1760–1766
139. Katsoyiannis IA, Zouboulis AI (2002) Removal of arsenic from contaminated water sources by sorption onto iron-oxide-coated polymeric materials. Water Res 36(20):5141–5155
140. Pan B, Li Z, Zhang Y, Xu J, Chen L, Dong H, Zhang W (2014) Acid and organic resistant nano-hydrated zirconium oxide (HZO)/polystyrene hybrid adsorbent for arsenic removal from water. Chem Eng J 248:290–296
141. Du Q, Zhang S, Pan B, Lv L, Zhang W, Zhang Q (2014) Effect of spatial distribution and aging of ZVI on the reactivity of resin-ZVI composites for arsenite removal. J Mater Sci 49 (20):7073–7079
142. Pan B, Han F, Nie G, Wu B, He K, Lu L (2014) New strategy to enhance phosphate removal from water by hydrous manganese oxide. Environ Sci Technol 48(9):5101–5107

Chapter 6
Polymer-Based Catalysts for Water Purification: Fundamentals to Applications

S. K. Shukla

Abstract The recent developments in the synthesis of polymer-based photocatalysts, photosensitizers, and hybrid photocatalysts along with their properties and potential applications in degradation of water pollutants have been presented. Polymer functions as photocatalysts, catalytic supports, and photosensitizers in pure as well as in composite form. The photocatalysts generate very reactive oxygen species (ROS), which efficiently oxidizes several pollutants such as dyes, pesticides, pharmaceuticals, and microorganism present in water. Polymeric and hybrid photocatalysts are especially well suited for removal of chemical compounds, which are present at low concentrations in water resources due to synergistic effect. The advantages for the use of photoactive polymeric are easy removal and long life, and control of the formation of secondary contamination is avoided.

6.1 Introduction

Exponential innovation in the synthesis, processing, and industrial application of polymer has compelled to call the current era as plastic age. The long-chain polymer along with synergic coherency in the different properties of polymer has additional advantages for the researchers to explore different dimension of polymers and their applications in agriculture, space engineering, electronics, water purification, and catalysis [1, 2]. The use of polymer in water purification is either membrane, adsorptions, or photocatalysts. The use of polymers in photocatalysis has several advantages like reduction in secondary pollutants, increase in surface area, collection and increase in efficiency [3, 4]. The other important features of polymer-based photocatalysis for water treatment are: (1) ambient operating temperature and pressure, (2) complete mineralization of parents and their intermediate compounds, and (3) low operating costs [5]. The role of polymers in photocatalysis

S. K. Shukla (✉)
Department of Polymer Science, Bhaskaracharya College of Applied Sciences, University of Delhi, New Delhi 110075, India
e-mail: sarojshukla2003@yahoo.co.in

R. Das (ed.), *Polymeric Materials for Clean Water*, Springer Series on Polymer and Composite Materials, https://doi.org/10.1007/978-3-030-00743-0_6

is either direct catalysts, support catalysts, or photosensitizer. Thus, both natural and synthetic polymer-based catalysts have been exponentially used for water purifications purpose [6]. Several polymer-based photocatalysts are used, and representatives are listed in Table 6.1.

Semiconductor-based photocatalysis is considered to be an attractive way for solving the worldwide energy shortage and environmental pollution issues. In this context, several polymeric semiconductor-based polymer composites have become a very hot research topic due to structure and band gap engineering [17, 18]. The composites of polymer with metal oxides also stabilize oxidation states of metal with unique chemical reactivity in photochemistry. The area is currently highly significant for removal of organic dyes, hydrocarbon, insecticides, and microorganism. The importance of area is also indicted by exponential increased in publication frequency, patent, and technical report. In light of above development, current chapter complies the resent development in the field of polymer-based photocatalysts and its application in water purification. The synthetic methods and important properties were also discussed in the significance of their use for water purification.

6.2 Synthetic Methods

Polymer-based photocatalysts can be prepared by direct and indirect methods using chemical, mechanical, photochemical, sono-chemical procedures. But uniform and homogeneous dispersion of nanoparticles in the polymer matrix is one of the major problems in fabrication of polymer nanocomposites. Thus, several innovations are carried out in this regard [19]. Shukla et al. have developed uniformly distributed SnO_2 in PANI from chemically functionalized monomers [20, 21]. The uniform incorporation of polymer nanostructures has attracted a lot of attention as photocatalyst for several applications [22, 23]. In this field, a recent experimental evidence of a visible-light-responsive photocatalytic activity of conjugated poly (diacetylene)-based polymers nanostructures, poly(diphenylbutadiyne) (PDPB) nanofibers for water de-pollution has reported with crucial role of the material structure at a nanometric scale [24]. The poly(3, 4-ethylenedioxythiophene) PEDOT nanostructures synthesized in soft templates via chemical oxidative polymerization demonstrate unprecedented photocatalytic activities for water treatment without the assistance of sacrificial reagents or noble metal co-catalysts and turn out to be better than TiO_2 as benchmark catalyst (Fig. 6.1). The PEDOT nanostructures showed a narrow band gap (E = 1.69 eV) with excellent ability to absorb light in visible and near-infrared region. This novel PEDOT-based photocatalysts are very stable with cycling and can be reused without appreciable loss of activity. Interestingly, hollow micrometric vesicular structures of PEDOT are not effective photocatalysts as compared to nanometric spindles suggesting size- and shape-dependent photocatalytic properties [25].

Table 6.1 Representative polymer photocatalysts, their synthesis and use for decontamination

Photocatalyst	Contaminants	Source	Fabrication method	Remark	References
ZnO nanorods/polybutylene terephthalate (PBT) fiber mats	Azo dye	UV radiation (320–390 nm) with 79 mW/ cm^2 energy flux	Vapor phase atomic layer deposition (ALD) and hydrothermal growth of ZnO nanorod crystals on a seed layer	Degradation ratio—90% of the dye within 2 h	[7]
Polyaniline (PANI) and ZnO nanoparticles	Methyl blue (MB) and eosin yellowish (EY) dye	Spectrum ranging from ultraviolet to visible (200–800 nm)	Impregnation of polymeric fiber using sol-gel process at ambient temperature	There is 77% MB dye degradation after 6 h upon ZnO/PANI. Similar results of degradation were obtained for EY-dye	[8]
CeO_2-ZnO-Polyvinylpyrrolidone (PVP)	Rhodamine B (RhB)	UV lamp (8 W) with 254 nm wavelength	The electrospinning technique was followed by thermal treatment to obtain CeO_2–ZnO-nanofibers$)_2$	98% decomposition applying for CeO_2–ZnO composite fibers better than individual	[9]
ZnO nanowires on polyethylene (PP)	Methylene blue (MB)	UV light source (6 W)	ZnO nanowire was grown on to the commercial available fibers by hydrothermal method	After 2.5 h of irradiation, ZnO/ polyethylene fibers degraded 83% of the MB	[10]
ZnO/SnO_2-Polyvinylpyrrolidone (PVP)	Rhodamine B (RhB)	High-pressure mercury lamp of 50 W	Hybrid method of sol-gel process and electrospinning technique	After 50 min, the degradation efficiency of RhB after 50 min equal to 75 and 35% for ZnO, and SnO_2. But the composite completely degraded in 30 min	[11]
Reduced graphene oxide/ titanium dioxide fiber (rGO/ TiO_2) and reduced graphene oxide/zinc oxide fiber (rGO/ ZnO) on polypropylene	Methylene blue (MB)	Halogen lamp (150 W)	The polypropylene (PP) was incorporated with reduced graphene oxide and metal oxide through hydrothermal approach	The composite shows more the 70% of MB degradation in 20 min but individual materials, degrades in 120 min 90%	[12]

(continued)

Table 6.1 (continued)

Photocatalyst	Contaminants	Source	Fabrication method	Remark	References
ZnAc/cellulose acetate (CA) composite nanofibers	Rhodamine B and phenol	Ultraviolet lamps (365 nm)	Electrospinning technique	Almost 100% of Rhodamine B and 85% phenol (after 24 h) was decomposed in the presence of TiO2/ZnO composite nanofibers under mild conditions	[13]
Titanium dioxide (TiO_2) immobilized in cellulose matrix	Phenol	UV (6 W) light at wavelength of 254 nm was used	Sol-gel method	The composite films exhibited high degradation ratio (90% after 2 h of irradiation)	[14]
Fe_3O_4/chitosan/TiO_2 nanocomposites	Methylene blue (MB)	UV light	solvents thermal reduction	The degradation rate of methyl blue was 93% after 30 min	[15]
Cellulose and activated carbon	Membrane	–	Solvent thermal casting	*E. coli*	[16]

Similarly, cross-linked polymers and cyclodextrins have been explored for removal of several polluting agents like chlorobenzene and chlorobiphenyl from water [26]. The formation of thin layers of photocatalyst in photomicroreactor is a challenging work due to the properties of catalyst and the microchannel material [27]. For example, the deposition of semiconductor materials on fluoropolymer-based microcapillary needs an economical method with less energy dependency. In this context, Colmenares et al. reported a novel method for depositing nanoparticles of TiO_2 on the inner walls of a hexafluoropropylene tetrafluoroethylene microtube under mild conditions employing ultrasound technique. The polymer surface was altered during ultrasonication, and it provides a site for deposition of catalyst like a thin layer of TiO_2 nanoparticles in the inner walls of the microtube. Further, the photocatalytic activity of the developed TiO_2-coated fluoropolymer-based microcapillary was evaluated for removal of phenol contaminants in water. The novel zincphthalocyanine-based conjugated microporous polymers with rigid-linker (α-ZnPc-CMP and β-ZnPc-CMP) were also synthesized by copolymerization of zinc phthalocyanine (ZnPc) and 4, 6-diaminoresorcinol dihydrochloride (DADHC). The α-ZnPc-CMP and β-ZnPc-CMP were used as a heterogeneous photocatalyst to degrade Rhodamine B (RhB) in aqueous solution. It is the recent report for use of MPc-based CMP-based heterogeneous photocatalysts for photodegradation of RhB. The highly ordered skeletal alignment and two-dimensional open-channel structure of α-ZnPc-CMP and β-ZnPc-CMP not only solve the aggregation of ZnPc and enhance its photocatalytic activity, but also facilitate the recycling and avoid the secondary pollution. The chemical structures and morphologies of α-ZnPc-CMP and β-ZnPc-CMP were well investigated by suitable techniques like Fourier transform infrared spectra (FT-IR), solid-state ^{13}C nuclear magnetic resonance (^{13}C NMR), scanning electron microscopy (SEM), N_2-sorption/desorption, and X-ray diffraction (XRD). The solubility experiments and thermogravimetric analysis (TGA) showed that it bears good chemical stability and recyclability. Furthermore, the photocatalytic tests indicated α-ZnPc-CMP and β-ZnPc-CMP have excellent photocatalytic performances for degradation of RhB (3 h, degraded 98 and 97.47%) in the presence of H_2O_2 under visible-light irradiation. All results reveal that α-ZnPc-CMP and β-ZnPc-CMP have great potential as photocatalysts on the degradation of organic dye contaminants. The possible reaction mechanism of α-ZnPc-CMP and β-ZnPc-CMP as photocatalysts for the degradation of RhB has been reported [28]. In another development, $H_3PW_{12}O_{40}$ (HPW)-containing polyimide (PI) hybrid composites (TPI) are prepared through in situ solid-state polymerization using HPW, melem, and pyromellitic dianhydride as precursors. The effect of HPW on the morphology, porosity, chemical structure, and optical and visible-light photocatalytic degradation efficiency of TPI composites are systematically investigated by various suitable analytical methods. By comparing the structure, property and photocatalytic activity of the TPI composites and the HPW-PI composites (prepared by the impregnation method), it inferred that the HPW can promote the formation of C–N bond in the five-membered imide rings between amines and anhydrides during the in situ solid-state condensation process. Consequently, the visible-light (λ > 400 nm)-based photocatalytic degradation

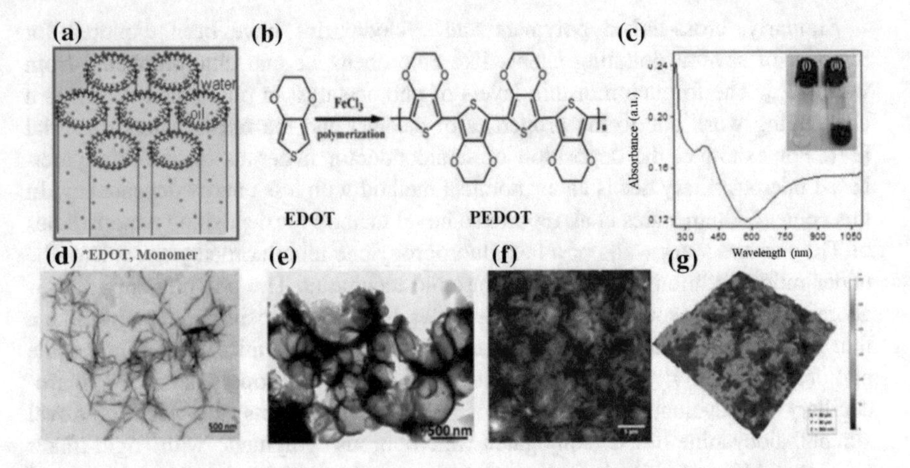

Fig. 6.1 Preparative steps and experimental evidence of PEDOT-based photocatalysts [25]. Copyright reserved with Nature Publishing Group

efficiency of imidacloprid on TPI composites is also enhanced than the pristine PI. The in situ solid-state condensation reaction also creates photogenerated electron–hole separation efficiency and visible-light utilization efficiency due to the introduction of HPW. The visible-light photocatalytic degradation rate constant k of 15% TPI composites prepared at 300 °C and 5% TPI composites prepared at 325 °C are about 10.33 and 2.42 times of the corresponding pristine PI, respectively [29]. Mechanochemical synthesis is another important method for preparation of polymer-based photocatalysts [30]. Banerjee et al. [31] prepared porous organic polymers POPs decorated with amide functionality using mechanochemical route. They have also compared their properties with the identical ones prepared by a conventional method. The prepared POPs were less surface area and show moderate adsorption properties but the presence of functional group shows remarkable stability in water and concentrated acids.

6.3 Properties

6.3.1 Photocatalysts

It is a class of materials responsible for acceleration of photoactive chemical reactions after absorption of light. It has wide range of applications such as photocatalytic removal of organic pollutants in water. The basic mechanism of a photochemical reaction is portrayed in Eq. 6.1:

$$\text{Pollutants} \xrightarrow[hv]{\text{Catalyst}} \text{Intermediate} \rightarrow CO_2 + H_2O \tag{6.1}$$

Generally, this reaction is comprised into different steps: (a) mass transfer of the organic contaminant; (b) adsorption of the organic contaminant(s) onto the photon activated catalysts; (c) photocatalysis reaction for the adsorbed phase; (d) desorption of the intermediate(s); (e) mass transfer of the intermediate(s) from the interface region to the bulk fluid. The absorption of photons with a specific energy allows for excitation of electrons from the valence band to the conduction band with generation of hole–electron pairs, which are responsible for progress in reaction. The reaction can be performed using different strategy; an exemplary case for photodegradation of water pollutants by a photochemical route is illustrated in Fig. 6.2.

The overall rate constants and efficiency depend on pollutants, catalysts, and irradiated radiations. Several catalysts like TiO_2, ZnO, SnO_2, and CeO_2 have been used in this regards. But basic limitations of these catalysts are atmospheric,

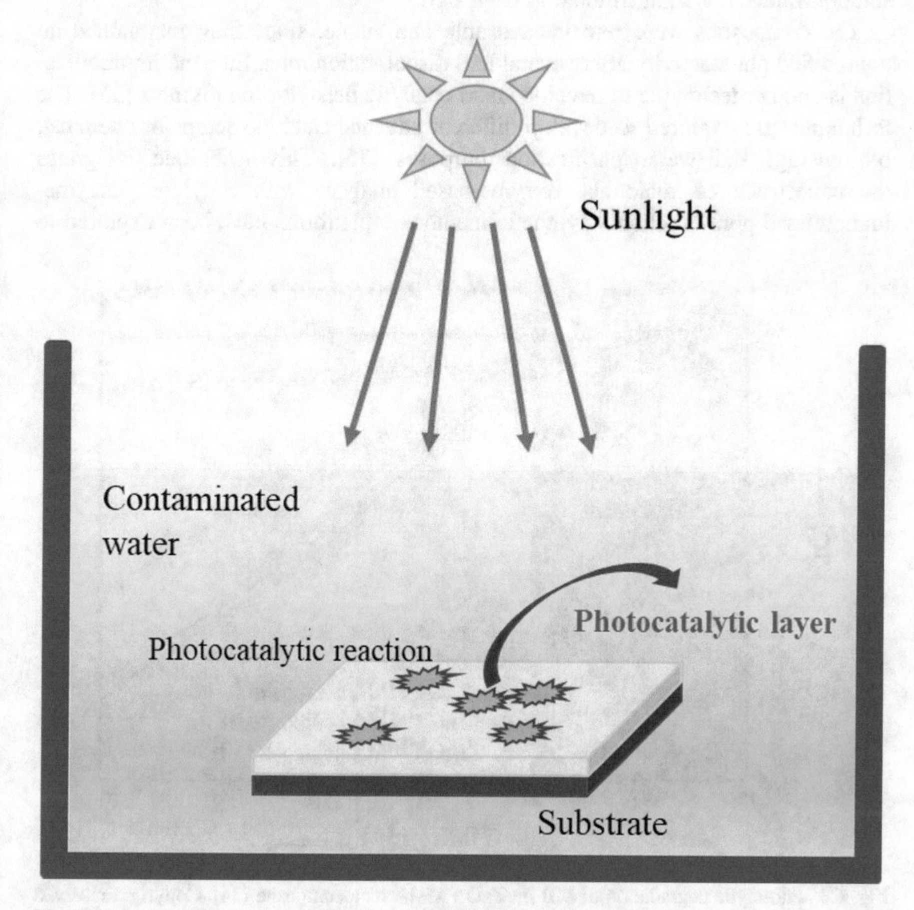

Fig. 6.2 An experimental setup for photocatalytic degradation [32]

chemical, and surface instability. Thus, synergism with polymers like conducting and coordination polymer has been explored to design efficient photocatalysts for different applications [33]. Based on the richness of metal-containing nodes and organic bridging linkers, as well as the controllability of the synthesis, it is easy to construct CPs with tailor able capacity to absorb light, thereby initiating desirable photocatalytic reactions for the degradation of organic pollutants. Several polymers have been investigated as photocatalysts for decomposition of organic dye molecules (organic pollutants) The incorporation of nanostructured photocatalysts like ZnO in a polymer, e.g., polymethyl methacrylate is a simple strategic to produce novel water purification systems [34]. This approach possesses the advantages of: (1) the presence of nanostructured photocatalyst; (2) the flexibility of polymer; (3) the immobilization of photocatalyst that avoids the recovery of the nanoparticles after the water treatment. The ZnO–polymer nanocomposites showed high photocatalytic performance and stability. The composite shows remarkable photocatalytic efficiency for the degradation of methylene blue (MB) dye and phenol in aqueous solution under UV light irradiation (Fig. 6.3).

The composites were remains reusable and stable, since they maintained an unmodified photoactivity after several MB discoloration runs. Enzyme immobilization is another technique to develop smart catalytic behavior in polymers [35]. The techniques are explored to develop different advance catalytic setup for chemical, bio-sensing, and water purification purposes [36]. This technique integrates the nanostructured materials, enzyme-based analysis with polymer. Enzyme-immobilized pore-functional synthetic membrane platforms have been explored to

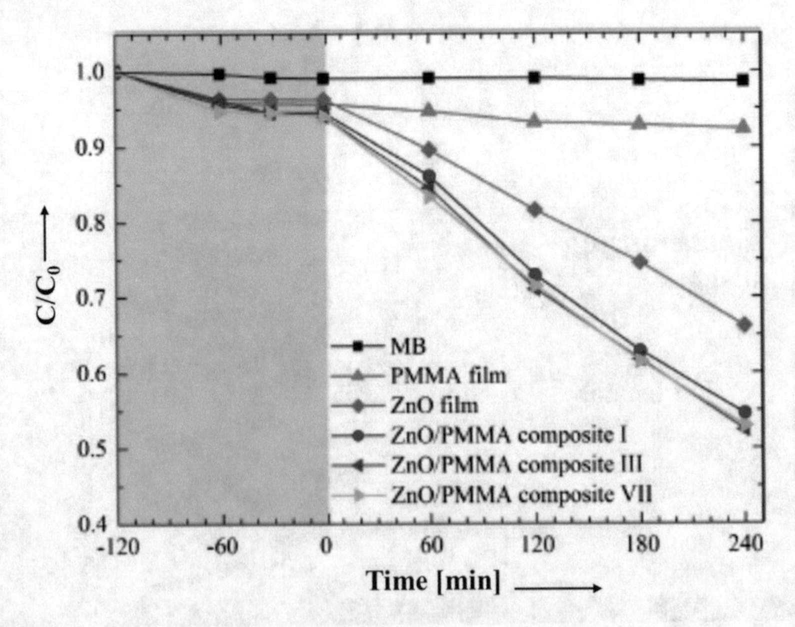

Fig. 6.3 Photolytic degradation of MB on ZnO/PMMA nanocomposite [34]. Copyright reserved with Nature Publishing Group

perform an environmentally important oxidative reaction for degradation toxic organic compound and detoxification of water without the addition of expensive or harmful chemicals [37]. The techniques have a promising future and industrial viability.

6.3.2 *Catalytic Supports*

The polymer matrix has found enormous applications as a support catalyst in order to improve efficiency for several applications. The use of polymer as supported catalyst increases the surface area, thermal and mechanical stability along with additional ability of reuse of catalysts [38, 39]. The polymer-supported photocatalysts are used in the organic synthesis, artificial photosynthesis, and water purifications. Polymeric supports possess different morphologies such as sheets [40], nanospheres [41], or nanoparticles [42]. All polymer materials contribute an increase in the photocatalytic activity of inorganic–organic materials. But interfacial contact surface of the hybrid photocatalyst also has a significant influence on their activity.

Wang et al. have demonstrated a support material by the use of TiO_2 nanospheres as the photoinitiator for photocatalytic surface-initiated polymerization and synthesis of various inorganic/polymer nanocomposites with well-defined structures. The preparative scheme is shown in Fig. 6.4 along with basic steps and conditions [43]. The excitation of TiO_2 by UV irradiation produces electrons and holes which drive the free radical polymerization near its surface. Thus, obtained core/shell composite nanospheres with eccentric or concentric structures can be tuned by controlling the surface compatibility between the polymer and the TiO_2. When highly porous TiO_2 nanospheres were employed as the photoinitiator, polymerization could disintegrate the mesoporous framework and give rise to nanocomposites with multiple TiO_2 nanoparticles evenly distributed in the polymer spheres. The sol-gel chemistry of titania is well-extendable to the coating of the polymers on many other substrates of interest such as silica and ZnS by simply premodifying their surface with a thin layer of titania. In addition, this strategy could be also easily applied to coating of different types of polymers such as polystyrene, poly(methyl methacrylate), and poly (*N*-isopropylacrylamide). This photocatalytic surface-initiated polymerization process could provide a platform for the synthesis of various inorganic/polymer hybrid nanocomposites for many interesting applications.

6.3.3 *Photosensitizers*

It is a class of molecule initiates a chemical change into another molecule during a photochemical process. Generally, it works after absorbing ultraviolet or visible region of electromagnetic radiation and transfer energy to desired reactants.

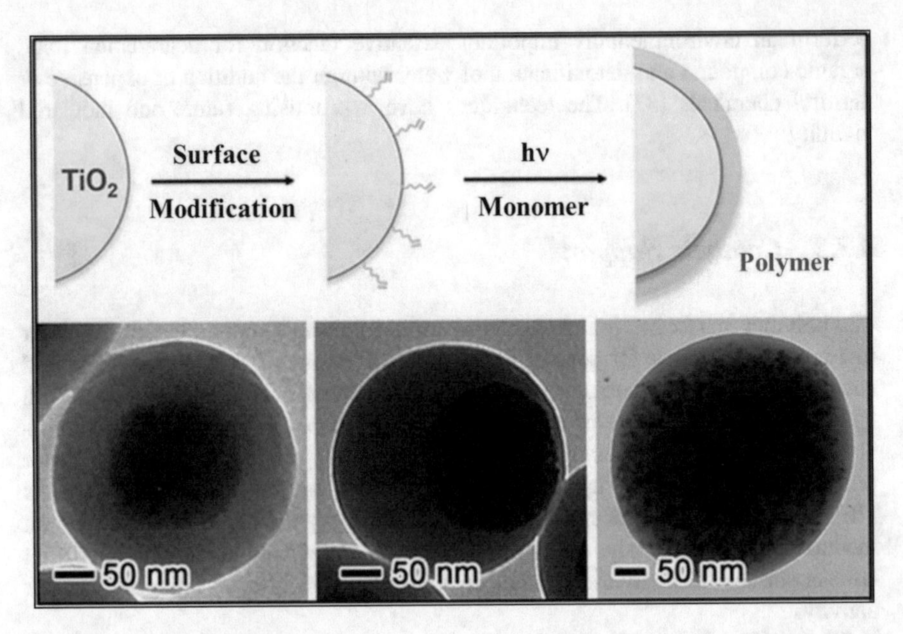

Fig. 6.4 Surface-initiated polymerization of inorganic/polymer nanocomposites [43]. Copyright reserved with American Chemical Society

The process is frequently used in several photochemical reactions like photodynamic therapy, photodegradation of water pollutants [44]. Several organic and inorganic photosensitizers are used for removal of different water pollutants like virus, bacteria, fungi, protozoa, dyes. 2, 1, 3-benzothiadiazole (BTZ)-based vinyl cross-linker was synthesized and copolymerized with large excesses of styrene using free radical polymerization to produce heterogeneous triplet photosensitizers in three distinct physical formats like gels, beads, and monoliths. These photosensitizers were explored for the generation of singlet oxygen ($^1O_2$) and for the aerobic hydroxylation of arylboronic acids via superoxide radical anion ($O_2^{\cdot -}$), thus obtained materials demonstrated good chemical and light stability. BTZ-containing beads and monoliths were exploited as photosensitizers in a commercial flow reactor, for production of $^1O_2$ through direct irradiation of sunlight. The conversion rate is comparable to the 420 nm LED module as a source of photons [45]. The photosensitizers (rose bengal (RB) and methylene blue (MB)) are when immobilized in polystyrene, which exhibits high antibacterial activity in a continuous regime. The photosensitizers were immobilized after dissolving it in a solvent like chloroform along with polystyrene. The mixed solution allows to evaporate solvent, the obtained films were used for constructing continuous-flow photoreactors for the removal of gram-positive *Staphylococcus aureus*, gram-negative *Escherichia coli* wastewater bacteria under illumination with visible white light using a luminescent lamp at a 1.8 mW cm^{-2} fluence rate. The bacterial concentration decreased by two to five orders of magnitude in separate reactors with either immobilized RB or MB, as well as in three reactors connected in series, which contained one of the photosensitizers. Bacterial eradication reached more than five orders of magnitude in two reactors connected in series, where the first reactor contained immobilized RB and the second contained immobilized MB [46]. Thus, it will be an alternative approach to the eradication of bacteria through photodynamic water treatment with the help of photosensitizers (PS). There are several advantages to using PS for wastewater disinfection. PS molecules are harmless to human beings and animals [47], and the disinfection process does not require an energetic impact since sunlight can be used for illumination. Furthermore, PS does not only inactivate bacteria, but also sewage bacteriophages [48] and promote photolysis of trace organic contaminants in the wastewater [49–52].

6.4 Applications

6.4.1 *Dyes*

Dyes are widely used materials in textiles, printing, rubber, cosmetics, plastics, and leather industries to color their products. The end use of product and the industrial affluent generates huge discharge of dyes in water bodies. This industrial discharge of dyes is posing serious threat to the water bodies due to toxic nature of organic dyes, which adversely affects plant, animal, and entire ecosystem [53]. Reactive dyes have been identified as problematic compounds because they are water soluble and presence in hydrolyzed form [54, 55]. The natural degradation of dye generates also several secondary pollutants. These pollutants are carcinogenic and biomagnified after consumption with aquatic animals. Further, majority of polymer materials exhibit a high level of resistance against ultraviolet irradiation and improved corrosion resistance. The chemical and environmental stability of the polymer hybrid catalyst depends, to a great extent on the functional support for removal of dyes. The drawbacks of natural and synthetic polymer-based photocatalytic materials and need special attention for the accessibility of synthetic polymeric materials derived from petroleum due to decreasing amounts of crude oil [56].

Both natural and synthetic polymer-based hybrid catalysts were explored for removal of dyes. A mechanistic design for dyes removal by floating photocatalysts was reported by Wang et al. (Fig. 6.5) [57]. Based on various types of properties polyethylene (PE), polypropylene (PP), polystyrene (PS), polyethylene terephthalate (PET), polyvinyl chloride (PVC), polyvinyl alcohol (PVA), polycarbonate (PC) etc. [58–63] have been reported as supports for photocatalytic in the literature. The first report to use polymer hybrid materials was made in 1995 by Tennakone [64]. Researcher has used titanium oxide with polyethylene films as support for the photocatalytic decomposition of phenol with a high degradation ratio (50% after 2.5 h of illumination). Further, research investigation on polypropylene non-woven with zinc oxide nanorods revealed that it exhibited excellent kind of photocatalytic activity along with high stability [65, 66]. Thus, different materials have been successfully explored as a photocatalysts for water treatment processes and filters [67]. Additionally, the synergetic effect between metal oxide and polymers allows for protection of the polypropylene fiber against surface cracks and limits photo-corrosion process of zinc oxide [68]. Similar photoactive hybrid materials based on polybutylene terephthalate (PBT) polymer fiber mats were used for photocatalytic dye degradation. The research finding has confirmed that the catalyst supported on the polymer mat could be reused without a particular recovery step [69]. It is also reported that fact that the combination of proper fabrication methods allows for better photocatalytic performance [70]. The other example of synthetic polymer hybrid used in water treatment are polyethersulfone or polyvinylidene fluoride membranes with various types of metal oxides (e.g., titanium, zinc, or chromium) displaying good antifouling performance, including photocatalysis, self-cleaning, and filterability properties [71]. The conjugated organic polymers(COPs) like polyaniline (PANI) [72], poly(pyrrole) (PPy) [73], polythiophene (PT) [74], poly-acetylene (PA) [75], poly(methyl methacrylate) (PMMA) [76], polythiopene (PT) [77], polyparaphenylene (PPP) [78], polyparaphenylenevenylene (PPV), poly (3, 4-ethylenedioxythiophene) (PEDOT) [78], or poly(Ophenylenediamine) (POPD)) based hybrids are also used [79]. The conjugated organic polymers are mostly p-type semiconductors with unique electrical and optical properties. Specifically, their high electron mobility or high photon absorption coefficient under visible spectra has attracted increasing interest for photocatalytic applications, e.g., degradation of pollutants or hydrogen generation by water splitting [80]. In terms of water treatment processes, another interesting perspective solution is offered by polymeric support [81]. In terms of water treatment processes, another interesting perspective solution offered by polymeric support is the possibility of fabricating a floatable photocatalyst, the concept of which is shown in Fig. 6.4. These kinds of materials are able to maximize illumination utilization and oxygenation processes of the photocatalyst by approaching the air/water interface. The result reveals the higher rates of radical formation and oxidation efficiencies [82]. Polymeric supports possess different morphologies such as sheets [83], nanospheres [84], or nanoparticles [85]. Overall, these morphologies of polymer materials contribute to an increase the photocatalytic activity of inorganic–organic materials. But contact surface area of the hybrid photocatalyst has a significant influence on their activity. Several examples of catalysts based on polymeric fibers with high photocatalytic activity are reported. Natural fibers are frequently used as a reinforcing composite for producing hybrid materials because they exhibit advantages, like recyclability and eco-friendliness, over their synthetic counterparts [86]. Additionally, natural fibers possess a higher volume fraction and larger loading capacity [87]. For these reasons, they are widely used to produce composite materials, especially in the field of photocatalysis. For instance, depositing titanium dioxide on cellulose fiber surface produces hybrid materials with a high degradation ratio of organic compounds like organic dye or phenolic contaminants. Yu et al. obtained cellulose-templated TiO_2/Ag nanosponge composites with improved

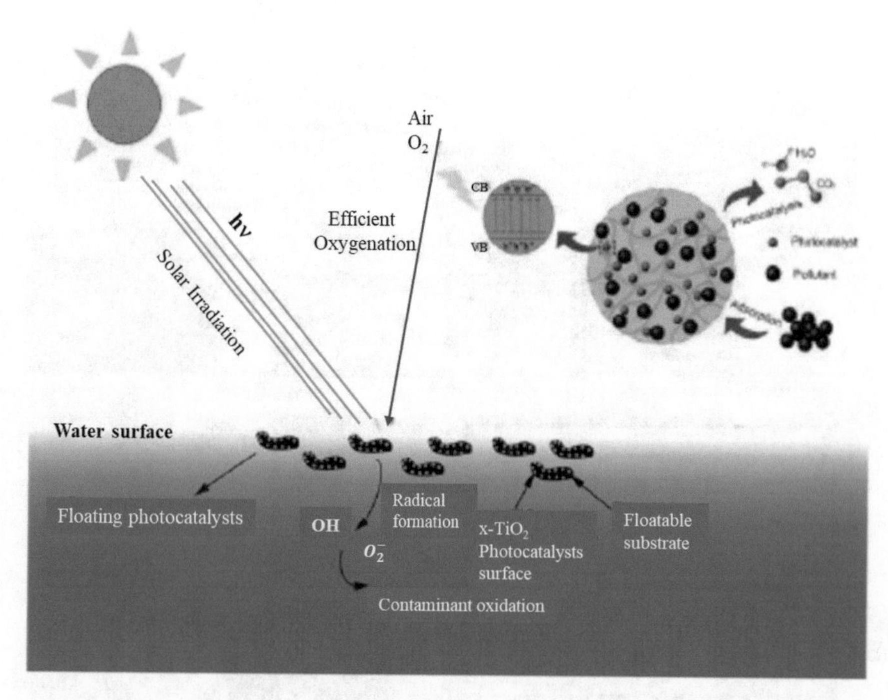

Fig. 6.5 Schematic representation of floating polymer catalysts [57]. Copyright reserved with MDPI, Basel, Switzerland

photocatalytic activities for the degradation of RhB [88]. The polymeric nanocomposite membranes with cellulose fibers can be also used for gas separation processes (e.g., hydrogen recovering, nitrogen generation, or carbon dioxide separation) [89, 90]. However, due to the fact that cellulose consists of monosaccharide units, it is hydrophilic and exhibits a rather poor interaction with most of the non-polar compound. Many efforts have been reported to obtain uniform dispersion of the fibers within the matrices. Furthermore, it is worth noting that plant fibers like cellulose possess relatively low processing temperatures. Novel photocatalyst membrane materials were successfully fabricated by an air jet spinning (AJS) technique from polyvinyl acetate (PVAc) solutions containing nanoparticles (NPs) of titanium dioxide (TiO_2). This innovative strategy is used for the production of composite nanofibers on stretching a solution of polymer through a high-speed compressed air jet. The technique rapidly allows to cover different substrates with TiO_2/PVAc interconnected nanofibers. Surprisingly, the diameters of the as-spun fibers were found to decrease with increasing amount of NPs. The results indicated that AJS-based PVAc-based fibrous membranes bears average fiber diameters of 505–901 nm have an apparent porosity of about 79–93% and a mean pore size of 1.58–5.12 μm. The embedding NPs onto the as-spun fibers resulted in increasing the tensile strength of the obtained composite fiber mats.

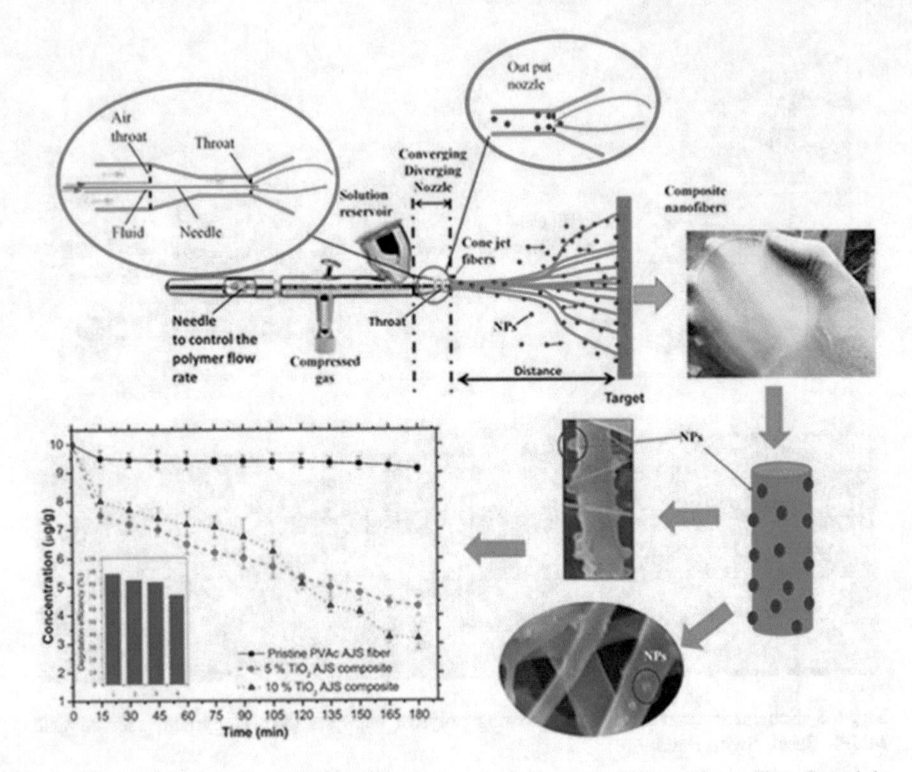

Fig. 6.6 Air jet technique for preparation of photocatalytic composites membrane [91]. Copyright reserved with American chemical society

The photodegradation property of TiO_2 membrane mats gives high efficiency in the decomposition of methylene blue dye. Thus, fiber spinning technique (Fig. 6.6) can provide the capacity to lace together a variety of types of polymers, fibers, and particles to produce interconnected fibers layer. This innovation opens the door for the innovation in nanocomposite mat that has great potential as efficient and economic water filter media with reusable photocatalyst [91].

6.4.2 *Hydrocarbons*

The hydrocarbons are the other common pollutants present in sediments of freshwater systems, urban water bodies, and sea water. The presence of hydrocarbon in water creates several negative properties like reduction in gas and heat exchange. Several hazardous reports are available on ill effects of hydrocarbon-based pollution. Generally, it is caused by accidents on oil platforms and ships used for hydrocarbon transport as well as also by discharging into the sea of water used to wash tanks of tanker vessels. Although hydrocarbons are simple organic substances with different chemical and physical properties, the hydrocarbon-based pollutant produces a waterproof film on water, which averts the exchange of oxygen between atmosphere and responsible for damages to marine flora and fauna. Environmental pollution caused by hydrocarbon is of great concern because petroleum hydrocarbons are toxic to all forms of life. The hydrocarbon is a complex mixture of saturated alkanes, branched alkanes, alkenes, napthenes (homo-cyclics and hetero-cyclics), aromatics (including aromatics containing hetero atoms like sulfur, oxygen, nitrogen, and other heavy metal complexes), naptheno-aromatics, large aromatic molecules like resins, asphaltenes, and hydrocarbon containing different functional groups like carboxylic acids, ethers [92, 93]. The occurrence of polycyclic aromatic hydrocarbons (PAHs) in surface water is frequently reported. In this regard, several remediation technologies are recommended, which are ex situ approach, steam-enhanced remediation chemical and photochemical oxidation. In this regard, several polymer catalysts have frequently used for their effective removal.

A bench-scale photodegradation experiment was reported for trace-level removal of anthracene, phenanthrene, and naphthalene by the use of stirred tank reactor and polymer-supported TiO_2 as a catalyst. The experimental design and conditions were best for the photocatalysis of the aqueous solutions and real samples in pH 9 and pH 7, and temperature 35 and 30 °C, respectively. Under the optimized conditions, the pollutants were completely degraded after 60 min of irradiation. The subproducts of the photocatalysis were identified through gas chromatography/mass spectrometry, and the fragmentation routes were discussed. The observed mean concentrations of PAHs in the polluted surface water and hospital wastewater were relatively high (3.9 ± 1.7 and 21.5 ± 2.8 $\mu g\ L^{-1}$ respectively). Thus, risk posed by the occurrence of PAHs in the surface water and hospital wastewater samples confirms the need of an efficient treatment system [94]. Bai et al. have removed approximately 80% PAHs (phenanthrene, fluoranthene, and benzo[a]pyrene by photocatalytic degradation over TiO_2 and graphene nanocomposite in 2 h). The chemical analysis of the degradation intermediate products indicates that the reaction is proceeded by formation of free radicals mechanism [95].

Another class of hydrocarbon-based pollutants is phenolics like cresols, which are harmful to both humans and our environment due to carcinogenic in nature. Their concentration in drinking water should not exceed 10 ppb. These pollutants are highly stable, soluble in water as well as volatile to some extent. They are also weakly adsorb on soil and tend to be bioaccumulated. These pollutants have been identified as stable, priority chemical toxicants by the United States Environmental Protection Agency [96] (USEPA, WHO) [97]. A modified poly (ether sulfone) (PES) by hydrophilic surface modifying macromolecules (LSMM) incorporated with oxygenated graphitic carbon nitride (OGCN) photocatalyst (PES/OGCN-LSMM) was successfully prepared as a hybrid photocatalytic membrane. The effect of solvent evaporation time during membrane fabrication was studied by focusing on the positioning of LSMM in order to provide the desirable properties of the PES/OGCN-LSMM hybrid membrane for phenol removal performance by photocatalytic and separation. The PES/OGCN-LSMM membranes exhibited a decreased value of contact angle as the solvent evaporation time increased. The special feature of LSMM tends to migrate upwards upon mixing, and the LSMM effectively assisted OGCN photocatalyst to the top layer of the membrane. It was found that the phenol reduction by rejection and photocatalytic tests was the highest at 5 min solvent evaporation time, while water flux was the lowest. The results revealed that the LSMM has indeed assisted the positioning of OGCN toward the top layer of the membrane and consequently increased the photocatalytic activity of the membrane on phenol [98].

An another photocatalytic reactor was developed by a novel flat sheet nanocomposite titanium dioxide (TiO_2)-halloysite nanotubes (HNTs)/polyvinylidene fluoride (PVDF) membrane as a photocatalytic separator (Fig. 6.7). The photocatalytic nanocomposite membrane played the roles for both degradation and separation for water. The hydrocarbon degradation and removal efficiency of the reactor was evaluated by gas chromatography mass spectroscopy (GC-MS) and reported that the reactor can remove 99.9% of hydrocarbons in 8 h. It is due to uniform distribution and high effectiveness of the TiO_2-HNTs photocatalyst in the PVDF polymer matrix. The TiO_2 leaching from the nanocomposite membrane during the membrane permeation was analyzed using flame atomic adsorption spectrophotometer (AAS) and reported that 1.0 ppb of TiO_2 leached in the permeate tank [99].

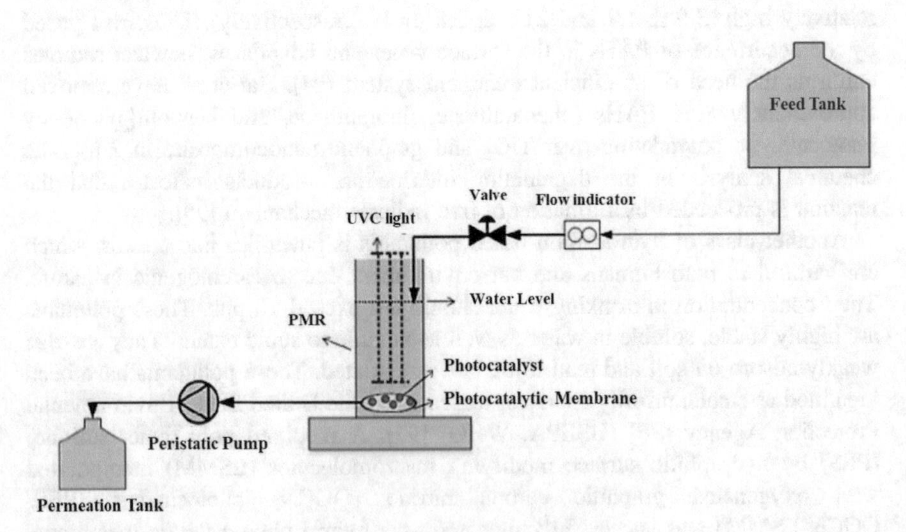

Fig. 6.7 Schematic of degradation and removal of hydrocarbon by photocatalytic membrane [99]. Copyright reserved with Royal Society of Chemistry, London

6.4.3 Pesticides and Pharmaceuticals

Accumulation of different pesticides and pharmaceuticals in the environment has attracted huge attention of the scientific community due their high consumption, low biodegradability, and toxic effects. Their presence in the environment is due to release during production, generation of domestic and hospital waste, human and animal excretion [100, 101]. The municipal wastewater treatment plants have been also identified as the source of this type of pollution in aquatic environments due the low removal efficiency to this class of compounds by conventional adopted treatment methods. Although pesticides, and agrochemical compounds in general, have been detected in water since the 1950s and 1960s, but in the last two decades, their use has risen dramatically all over the world. According to United Nations report, out of total pesticides used in agriculture only less than 1% actually reaches the crops. The remainder ends up to contaminate land, air, and water. It is reported that the pesticide-based water pollution is one of the greatest environmental problems, which have widespread ecological consequences. Another biggest source of pesticides is the uncontrolled dumping of empty pesticide containers. These containers have a very small amount of pesticide but being dumped in large quantities (millions per year), and it has become one of the most important sources of pollution in areas [102]. A vital solution to this problem will be collection and transportation of plastic bottles to a recycling plant after proper cleaning for a posterior reuse but not in practice. As a consequence, it is necessary to develop simple, inexpensive, and accessible technologies for in situ treatment of pesticides from contaminated waters. Traditionally, it is removed using granular or powdered activated carbon, nanofiltration, ozonation, and isolation of specific bacterial cultures with their inherent limitations in applicability, effectiveness, and costs. The problems associated with the disposal of these chemicals to the environment fit quite well with a polymer-based photocatalytic treatment. Solar photocatalysis has been shown to be very efficient in the degradation of these compounds. A nanoporous polymeric crystalline TiO_2 composite (TiO_2/PDVB-MA) has been successfully synthesized through an in situ synthesis method using divinylbenzene (DVB), methacrylic acid (MA), and tetrabutyl titanate. The TiO_2/PDVB-MA composite was explored used as a photocatalyst for degradation of Rhodamine B (RhB), bisphenol A, and 2, 4, 6-trichlorophenol under irradiation of visible light. Interestingly, excellent photocatalytic performance of the composite was found for RhB and bisphenol A, which is may be due to the synergism between TiO_2 and PDVB-MA. The developed TiO_2/PDVB-MA composite has been recycled at four times for the removal of RhB. It reveals that the composite is a promising photocatalyst to degrade the organic pollutants under visible-light irradiation [103]. Antibiotics are another class of organic pollutants in aquatic environments; however, the contribution of antibiotic exposure in human is not well-explored [104]. Some of antibiotics detected in water bodies are fluoroquinolones, sulfonamides, lincomycin, tetracyclines, and macrolides. Their presence in drinking water is of concern due to the unknown health effects of chronic low-level exposure to antibiotics over a lifetime. Several advanced treatment systems, including membrane filtration, granular activated carbon, and advanced oxidation processes, have been used for the effective removal of antibiotic. Karaolia et al. have exhaustively reported on (i) the removal of the antibiotics sulfamethoxazole (SMX), erythromycin (ERY) and clarithromycin (CLA); (ii) the inactivation of the total and antibiotic-resistant *E. coli* along with their regrowth potential after treatment; (iii) the removal of the total genomic DNA content; and (iv) the removal of selected antibiotic resistance genes (ARGs). In this regard, TiO_2 and polymer photocatalyst have been used under solar radiation in real urban wastewaters. TiO_2-reduced graphene oxide (TiO_2-rGO) composite photocatalysts were synthesized by two ex situ synthesis methods, namely hydrothermal (HD) treatment and photocatalytic (PH) treatment, starting from graphene oxide and TiO_2. The potential of the synthesized TiO2-rGO composites for the removal of the above-mentioned antibiotic-related micro contaminants was compared to the efficiency shown by pristine TiO_2 under simulated solar radiation for real urban wastewater effluents treated employing a membrane bioreactor. The results indiacte that TiO_2-rGO-PH was more efficient in the photocatalytic degradation of ERY by 84 ± 2%, CLA by 86 ± 5%, and degradation of SMX by 87 ± 4%. The degradation efficiency was higher than TiO_2 and degradation completed in more than 180 min. The treatments were suitable for the complete inactivation and complete absence of post-treatment regrowth of *E. coli* bacteria (<LOD) even 24 h after the end of the treatment. The least amount of regrowth at all experimental times was observed in the presence of TiO_2-rGO-HD. Further, the synthesized graphene-based photocatalysts were successfully removed ampC and significantly reduced ecfX, but sul1 and 23S rRNA for enterococci sequences were found to be persistent throughout treatment with all catalyst types. However, total DNA concentration remained stable throughout the photocatalytic treatment (4.2–4.8 ng μL^{-1}) [105]. The photocatalysts were capable of removing the target antibiotics in real wastewater effluents under simulated solar radiation. TiO_2/Biocidal polymer nanoparticles were synthesized by surface-initiated photopolymerization using TiO_2 nanoparticles as photoinitiators. Thus, obtained nanocomposites exhibited excellent antimicrobial properties under both dark and UV irradiation. Novel biocidal polymer-functionalized TiO_2 nanoparticle was prepared by surface-initiated photopolymerization using titania as an initiator. Vinyl monomer mixtures of nontoxic secondary amine-containing biocidal 2-(*tert*-butylamino)ethyl methacrylate and antifouling ethylene glycol dimethacrylate were used for the antimicrobial polymer shell. The synthesized TiO_2/poly[2-(*tert*-butylamino)ethyl methacrylate-*co*-ethylene glycol dimethacrylate] core/shell nanoparticles had enhanced photocatalytic antibacterial properties than pristine TiO_2 nanoparticles due to the combined antibacterial activities of light-driven anti-infective TiO_2 core and biocidal polymer shell. In the dark condition, the TiO_2/biocidal polymer nanoparticles exhibited high antimicrobial efficiency (95.7%) against gram-positive *S. aureus*. But after UV irradiation, the TiO_2/biocidal polymer showed improved inhibition of bacterial growth against

gram-negative *E. coli* and gram-positive *S. aureus* in comparison to the pristine TiO_2 nanoparticles. [106]

Microbial water pollutions are causing several serious diseases such as cholera due to presence of microorganisms in water. It is observed that microbe-contaminated water serves as disease vehicles. These diseases usually adversely affect the health of people in poorer countries due to lack facilities to treat polluted water [107]. The important microbial pollutants are bacteria, viruses, and protozoa. Increasing interest in controlling water-borne pathogens in water resources evidenced by a large number of recent publications for synthesis of synthesize knowledge from multiple fields covering comparative aspects of pathogen contamination, and unify them in a single place in order to present and address the problem as a whole [108]. Several technologies such as membrane filtration, inactivation, and photolytic killing are explored for removal of pathogen. A photoreactor for continuous inhibition of suspended bacteria was designed by Nisnevitch (Fig. 6.8). The reactor is based on one to three reservoirs: Shallow reservoirs were bottom-coated with immobilized PS, but control reservoirs were coated with polystyrene without PS Immobilized PS exhibited good photokilling abilities on a model of Gram-negative and Gram-positive bacteria (*E. coli* and *Enterococcus faecalis*, respectively) and wastewater fecal bacteria in a batch and in a continuous regime. The immobilized PS demonstrated higher stability and resistance against photobleaching than free PS, and maintained at least some of their antibacterial activity after storage for several months in the dark. Numerous studies have been undertaken to nurture the knowledge and understanding of this process [109]. Early insights into the bactericidal mechanism action of TiO_2 photocatalysis were discussed by Matsunaga et al. [110]. They have demonstrated the direct oxidation of intracellular Coenzyme A in the bacteria Lactobacillus acidophilus and *E. coli* and in the yeast *Saccharomyces cerevisia*. This resulted in the inhibition of respiratory activity and eventual cause the death of the cell. After that Saito et al. [111] showed that cell death was accompanied by a rapid leakage of potassium ions along with the slow release of bacterial protein and RNA from *Streptococcus sobrinus* AHT. The inference from transmission electron micrographs of treated *S. sobrinus* AHT showed that cell death was due to a significant disorder of cell membranes as well as cell wall decomposition. Further, the evidence for cell membrane involvement in the photocatalytic killing process was also reported by other groups [112, 113]. The usefulness of photocatalysis for the disinfection of water has been explored for wide range of microorganisms, i.e., bacteria, virus [114], viruses [115], fungi [116], and protozoa [117].

Furthermore, formulation affects the degradation process and, unfortunately, very little information on the effects of adjutants in photocatalysis degradation is available. Despite recent efforts in modernization of water treatment facilities, the problem of access to healthy drinking water for hundreds of millions of people has still not been solved. A water filter based on Cu-coated nanofibrillated cellulose

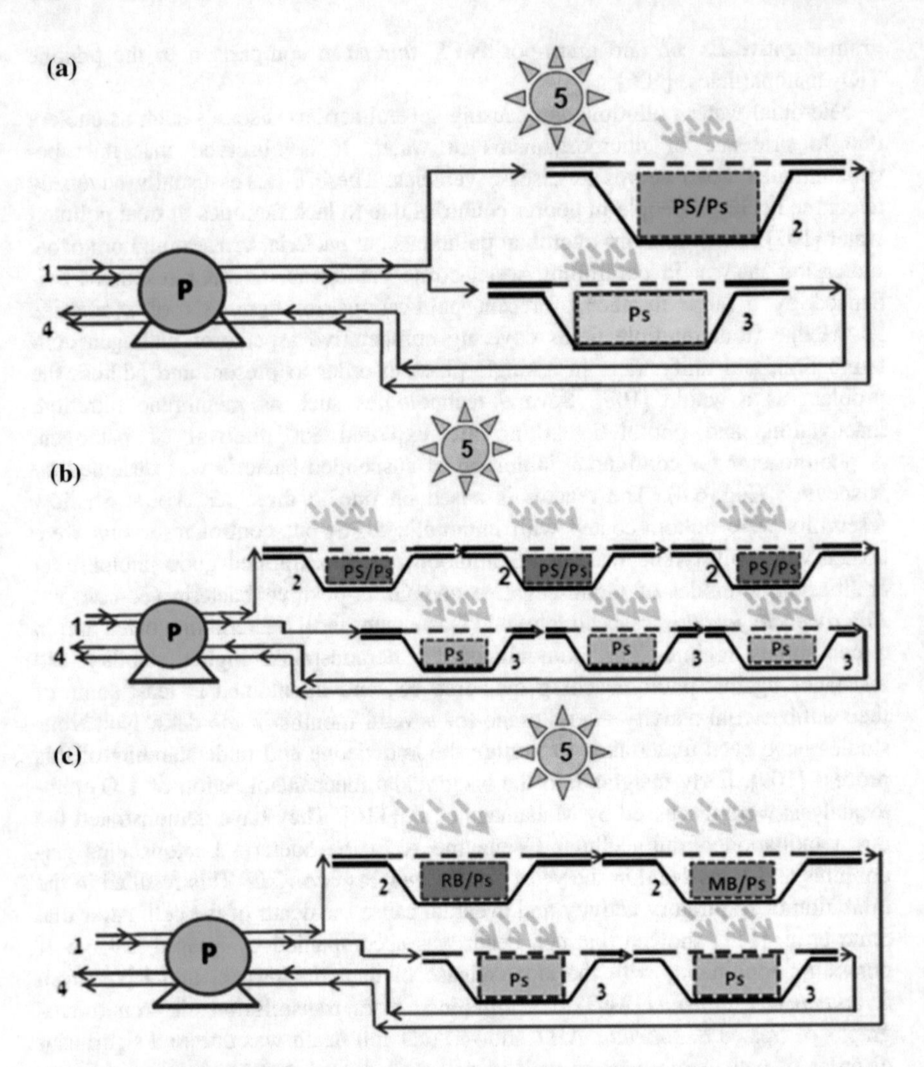

Fig. 6.8 A photoreactor scheme for continuous eradication of bacteria [46]. Copyright reserved with MDPI, Basel, Switzerland

with controlled porosity was prepared by the "paper-making" method. This optimized the proper mass and ratio of functionalized and pure nanofibrillated cellulose for the preparation of the filter. The filter material was tested in batch experiments and the fixed filters in flow experiments. The fabricated Cu-coated nanofibrillated cellulose filters were characterized for different properties, morphology, and chemical structure by appropriate techniques. The results revealed that the fixation of cellulose nanofibers plays a significant role in the degree of virus retention, and it greatly enhances the efficiency of the filtration. The functionalized water filters were able to achieve virus retention of at least 5 magnitudes (5Log) at three different pH values: 5.0, 7.5, and 9 [118].

6.5 Summary

The multifunctionality in polymer makes it suitable for photocatalytic purification of water as catalysts, support, and photosensitizers. The both natural and synthetic polymers have been explored in context employing direct and direct methods. The polymer hybrids exhibit significantly better photocatalytic properties than the separate components, due to the synergistic effect coming from the intrinsic properties of a photoactive semiconductor and polymers. Several key advantages can be expected from polymeric support, such as: (a) an increase of the specific surface area which consequently allows for adsorption of higher amounts of target pollutants [119–121] and (b) an improvement of the photocatalytic performance by promoting reduction of the charge carriers recombination and prolongation of the photoelectron lifetime [122]. In this chapter, the highlighted benefits and drawbacks of natural and synthetic polymer-based photocatalytic for removal water pollutants like dyes, hydrocarbon, and microbes. Currently, the scientific world indicates that polymer materials are the key promising components of the next generation of photocatalytic hybrid materials for water treatment [123, 124]. However, there are still some limitations (Table 6.2) in this which are still need to intensively studied.

Table 6.2 Comparison between significant properties of synthetic and biopolymers

Properties and characteristics	Synthetic polymers	Biopolymers
Availability	Decreasing	High
Physicochemical resistance	High	Low
Thermal stability	High	Low
Large scale applications	Possible	Difficult
Environmental friendly	No	Yes
Cost of production	Low	High
Sustainability	Low	High

References and Future Readings

1. Pichat P, Oills D (2013) Photocatalytic treatment of water: Irradiance influences. In: Pichat P. (ed) Photocatalysis and water purification: From fundamentals to recent applications, 1st edn. Wiley-VCH Verlag GmbH & Co., KGaA, Germany, pp 311–333
2. Korina E, Stoilova O, Manolova N, Rashkov I (2018) Polymer fibers with magnetic core decorated with titanium dioxide prospective for photocatalytic water treatment. J Environ Chem Eng 6:2075–2084
3. a. Shannon MA, Bohn PW, Elimelech M, Georgiadis JG, Marin BJ, Mayes AM (2017) Science and technology for water purification in the coming decades. Nature 5(2):301–310. b. Mamaghani AH, Haghighat F, Lee C-S (2017) Photocatalytic oxidation technology for indoor environment air purification: the state-of-the-art. Appl Catal B 203:247–269
4. Borges ME, García DM, Hernández T, Ruiz-Morales JC, Esparza P (2015) Supported photocatalyst for removal of emerging contaminants from wastewater in a continuous packed-bed photoreactor configuration. Catalysts 5:77–87
5. Chong MN, Jin B, Chow CWK, Saint C (2010) Recent developments in photocatalytic water treatment technology: a review. Water Res 44:2997–3027
6. Colmenares JC, Kuna E (2017) Photoactive hybrid catalysts based on natural and synthetic polymers: a comparative overview. Molecules 22:790
7. Gong B, Peng Q, Na JS, Parsons GN (2011) Highly active photocatalytic ZnO nanocrystalline rods supported on polymer fiber mats: synthesis using atomic layer deposition and hydrothermal crystal growth. Appl Catal A Gen 407:211–216
8. Moafi HF, Shojaie AF, Zanjanchi MA (2011) Semiconductor-assisted self-cleaning polymeric fibers based on zinc oxide nanoparticles. J Appl Polym Sci 121:3641–3650
9. Li C, Chen R, Zhang X, Shu S, Xiong J, Zheng Y, Dong W (2011) Electrospinning of CeO2–ZnO composite nanofibers and their photocatalytic property. Mater Lett 65:1327–1330
10. Baruah S, Thanachayanont C, Dutta J (2008) Growth of ZnO nanowires on nonwoven polyethylene fibers. Sci Technol Adv Mater 9:025009
11. Zhang Z, Shao C, Li X, Zhang L, Xue H, Wang C, Liu Y (2010) Electrospun nanofibers of ZnO/SnO2 heterojunction with high photocatalytic activity. J Phys Chem C 114:7920–7925
12. Ariffin SN, Lima HN, Jumeri FA, Zobir M, Abdullah AH, Ahmad M, Ibrahim NA, Huang NM, Teo PS, Muthoosamy K et al (2014) Modification of polypropylene filter with metal oxide and reduced graphene oxide for water treatment. Ceram Int 40:6927–6936
13. Liu H, Yang J, Liang J, Huang J, Tang C (2008) ZnO nanofiber and nanoparticle synthesized through electrospinning and their photocatalytic activity under visible light. J Am Chem Soc 91:1287–1291
14. Zeng J, Liu S, Cai J, Zhang L (2010) TiO_2 immobilized in cellulose matrix for photocatalytic degradation of phenol underweak UV light irradiation. J Phys Chem C 114:7806–7811
15. Choia C, Hwanga KJ, Kimb YJ, Kimb G, Parkc JY, Sungho J (2016) Rice-straw-derived hybrid TiO_2–SiO_2 structures with enhanced photocatalytic properties for removal of hazardous dye in aqueous solutions. Nano Energy 20:76–83
16. Hassan M, Abou-Zeid R, Hassan E, Berglund L, Aitomäki Y, Oksman K (2017) Membranes based on cellulose nanofibers and activated carbon for removal of *Escherichia coli* bacteria from water. Polymers 9(8):335
17. Cao S, Low J, Yu J, Jaron M (2015) Polymeric photocatalysts based on graphitic carbon nitride. Adv Mater 27(13):2150–2176
18. Zare IN, Motahari A, Sillanpää M (2018) Nanoadsorbents based on conducting polymer nanocomposites with main focus on polyaniline and its derivatives for removal of heavy metal ions/ dyes: a review. Environ Res 162:173–195
19. Heng PH, Sun Y, Liu X (2018) In situ polymerization synthesis of Z-scheme tungsten trioxide/polyimide photocatalyst with enhanced visible-light photocatalytic activity. Appl Surf Sci 428:1130–1140

20. Shukla SK, Shukla Sudheesh K, Govender Penny P, Agorku Eric S (2016) A resistive type humidity sensor based on crystalline tin oxide nanoparticles encapsulated in polyaniline matrix. Microchim Acta 183:573–580
21. Pandey N, Shukla SK, Singh NB (2017) Water purification by polymer nanocomposites: an overview. Nanocomposites 3(2):47–66
22. Muktha B, Madras G, Gururow TN, Scherf U, Patil S (2007) Conjugated polymers for photocatalysis. J Phys Chem B 111(28):7994–7998
23. Ghosh S et al (2015) Conducting polymer nanostructures for photocatalysis under visible light. Nat Mater 14:505–511
24. Yin Z, Zheng Q (2012) Controlled synthesis and energy applications of one-dimensional conducting polymer nanostructures: an overview. Adv Energy Mater 2:179–218
25. Ghosh S, Kouame NA, Remita S, Ramos L, Goubard F, Aubert P-H, Dazzil A, Deniset-Besseau A, Remita H (2015) Visible-light active conducting polymer nanostructures with superior photocatalytic activity. Sci Rep 5:18002
26. United States (12) Patent Application Publication (10) Pub. No.: US 2005/0154198A1
27. Colmenares JC, Nair V, Kuna E, Łomot D (2018) Development of photocatalyst coated fluoropolymer based microreactor using ultrasound for water remediation. Ultrason Sonochem 41:297–302
28. Cai L, Li Y, Li Y, Wang H, Yu Y, Liu Y, Duan Q (2018) Synthesis of zincphthalocyanine-based conjugated microporous polymers with rigid-linker as novel and green heterogeneous photocatalysts. J Hazard Mater 348:47–55
29. Meng P, Heng H, Sun Y, Huang J, Yang J, Liu X (2018) Positive effects of phosphotungstic acid on the in-situ solid-state polymerization and visible light photocatalytic activity of polyimide-based photocatalyst. Appl Catal B 226:487–498
30. Do J-L, Friščic T (2017) Mechanochemistry: a force of synthesis. ACS Cent Sci 3:13–19
31. Rajput L, Banerjee R (2014) Mechanochemical synthesis of amide functionalized porous organic polymers. Cryst Growth Des 14(6):2729–2732
32. Gupta A, Saurav JR, Bhattacharya S (2015) Solar light based degradation of organic pollutants using ZnO nanobrushes for water filtration. RSC Adv 5:71472
33. Zhou Y-B, Zhan Z-P (2018) Conjugated microporous polymers for heterogeneous catalysis. Chem Asian J 13:9–19
34. Mauro AD, Cantarella M, Nicotra G, Pellegrino G, Gulino A, Brundo MV, Privitera V, Impellizzeri G (2017) Novel synthesis of ZnO/PMMA nanocomposites for photocatalytic applications. Sci Rep 7:40895
35. Cirillo G, Nicoletta FP, Curcio M, Spizzirri UG, Picci N, Iemma F (2014) Enzyme immobilization on smart polymers: catalysis on demand. React Funct Polym 83:62–69
36. Barbosa O, Torres R, Ortiz C, Berenguer-Murcia Á, Rodrigues RC, Fernandez-Lafuente R (2013) Heterofunctional supports in enzyme immobilization: from traditional immobilization protocols to opportunities in tuning enzyme properties. Biomacromol 14(8):2433–2462
37. Lewis SR, Datta S, Gui M, Coker EL, Huggins FE, Daunert S, Bachas L, Bhattacharyya D (2011) Reactive nanostructured membranes for water purification. PNAS 108:8577–8582
38. Lu J, Toy PH (2009) Organic polymer supports for synthesis and for reagent and catalyst immobilization. Chem Rev 109(2):815–838
39. Benaglia M, Puglisi A, Cozzi F (2003) Polymer-supported organic catalysts. Chem Rev 103(9):3401–3430
40. Naskar S, Pillay AS, Chanda M (1998) Photocatalytic degradation of organic dyes in aqueous solution with TiO_2 nanoparticles immobilized on foamed polyethylene sheet. J Photochem Photobiol A Chem 3:257–264
41. Jin L, Wu H, Morbidelli M (2015) Synthesis of water-based dispersions of polymer/TiO_2 hybrid nanospheres. Nanomaterials 5:1454–1468
42. Nabid MR, Golbabaee M, Moghaddam AB, Dinarvand R, Sedghi R (2008) Polyaniline/TiO_2 nanocomposite: enzymatic synthesis and electrochemical properties. Int J Electrochem Sci 3:1117–1126

43. Wang X, Lu Q, Wang X, Joo J, Dahl M, Liu B, Gao C, Yin Y (2016) Photocatalytic surface-initiated polymerization on TiO_2 toward well-defined composite nanostructures. ACS Appl Mater Interfaces 8(1):538–554
44. Thandu M, Comuzzi C, Goi D (2015) Phototreatment of water by organic photosensitizers and comparison with inorganic semiconductors. Int J Photoenergy, Article ID 521367, 22 pages
45. Tobin JM, McCabe TJD, Prentice AW, Holzer S, Lloyd GO, Paterson MJ, Arrighi V, Cormack PAG, Vilela F (2017) Polymer-supported photosensitizers for oxidative organic transformations in flow and under visible light irradiation. ACS Catal 7:4602–4612
46. Valkov A, Nakonechny F, Nisnevitch M (2014) Polymer-immobilized photosensitizers for continuous eradication of bacteria. Int J Mol Sci 15:14984–14996
47. Macdonald IJ, Dougherty TJ (2001) Basic principles of photodynamic therapy. J Porphyr Phthalocyanines 5:105–129
48. Costa L, Carvalho CMB, Faustino MAF, Neves MG, Tomé JPC, Tomé AC, Cavaleiro JAS, Cunha A, Almeida A (2010) Sewage bacteriophage inactivation by cationic porphyrins: influence of light parameters. Photochem Photobiol Sci 9:1126–1133
49. Jasper JT, Sedlak DL (2013) Phototransformation of wastewater-derived trace organic contaminants in open-water unit process treatment wetlands. Environ Sci Technol 47:10781–10790
50. Hussein FH (2012) Comparison between solar and artificial photocatalytic decolorization of textile industrial wastewater. Int J Photoenergy 2012:793648
51. DeRosa MC, Crutchley RJ (2002) Photosensitized singlet oxygen and its applications. Coord Chem Rev 233–234:351–371
52. Nowakowska M, Kępczyński M (1998) Polymeric photosensitizers 2. Photosensitized oxidation of phenol in aqueous solution. J Photochem Photobiol 116:251–256
53. Kandisa RV, Narayana Saibaba KV, Shaik KB, Gopinath R (2016) Dye removal by adsorption: a review. J Bioremediat Biodegrad 7(6)
54. Tunc O, Hacer T, Aksu Z (2009) Potential use of cotton plant wastes for the removal of Remazol Black B reactive dye. J Hazard Mater 163:187
55. Renganathan S, Kalpana J, Kumar M, Velan M (2009) Equilibrium and Kinetic Studies on the Removal of Reactive Red 2 Dye from an Aqueous Solution Using a Positively Charged Functional Group of the *Nymphaea rubra* Biosorbent. CLEAN Soil Air Water 37:901
56. Valkov A, Nakonechny F, Nisnevitch M (2014) Polymer-immobilized photosensitizers for continuous eradication of bacteria. Int J Mol Sci 15:14984–14996
57. Wang RX, Wang W, Wang X, Zhang J, Gu Z, Zhou L, Zhao J (2015) Enhanced visible light photocatalytic activity of a floating photocatalyst based on B-N-codoped TiO_2 grafted on expanded perlite. RSc Adv 5:41385–41392
58. Yu Z, Mielczarski E, Mielczarski J, Laub C, Buffat P, Klehm U, Albers P, Lee K, Kulike A, Kiwi-Minerska L (2007) Preparation, stabilization and characterization of TiO_2 on thin polyethylene films (LDPE). Photocatalytic applications. Water Res 41:862–874
59. Ma S, Meng J, Li J, Zhang Y, Ni L (2014) Synthesis of catalytic polypropylene membranes enabling visible-light-driven photocatalytic degradation of dyes in water. J Membr Sci 453:221–229
60. Zan L, Tian L, Liu Z, Peng Z (2004) A new polystyrene–TiO_2 nanocomposite film and its photocatalytic degradation. Appl Catal A Gen 264:237–242
61. Taylor DM, Lewis TJ (1971) Electrical conduction in polyethylene terephthalate and polyethylene films. J Phys D Appl Phys 4:1346–1354
62. Wang D, Shi L, Luo Q, Li X, An J (2012) An efficient visible light photocatalyst prepared from TiO_2 and polyvinyl chloride. J Mater Sci 47:2136–2145
63. Araújo VD, Tranquilin RL, Motta FV, Paskocimas CA, Bernardi MIB, Cavalcante LS, Andres JS, Longo E, Bomio MRD (2014) Effect of polyvinyl alcohol on the shape, photoluminescence and photocatalytic properties of $PbMoO_4$ microcrystals. Mater Sci Semicond Process 26:425–430

64. Tennakone K, Tilakaratne CTK, Kottegoda IRM (1995) Photocatalytic degradation of organic contaminants in water with TiO_2 supported on polyethene films. J Photochem Photobiol A Chem 87:177–179
65. Ariffin SN, Lima HN, Jumeri FA, Zobir M, Abdullah AH, Ahmad M, Ibrahim NA, Huang NM, Teo PS, Muthoosamy K (2014) Modification of polypropylene filter with metal oxide and reduced graphene oxide for water treatment. Ceram Int 40:6927–6936
66. Li M, Li G, Fan Y, Jiang J, Ding Q, Dai X, Mai K (2014) Effect of nano-ZnO-supported 13X zeolite on photo-oxidation degradation and antimicrobial properties of polypropylene random copolymer. Polym Bull 71:2981–2997
67. Colmenares JC, Kuna E, Jakubiak S, Michalski J, Kurzydłowski K (2015) Polypropylene nonwoven filter with nanosized ZnO rods: promising hybrid photocatalyst for water purification. Appl Catal B Environ 170–171:273–282
68. Sakthivel S, Neppolian B, Shankar MV, Arabindoo B, Palanichamy M, Murugesan V (2003) Solar photocatalytic degradation of azo dye: comparison of photocatalytic efficiency of ZnO and TiO_2. Sol Energy Mater Sol Cells 77:65–82
69. Mamaghani AH, Haghighat F, Lee C-S (2017) Photocatalytic oxidation technology for indoor environment air purification: the state-of-the-art. Appl Catal B 203:247–269
70. Gong B, Peng Q, Na JS, Parsons GN (2011) Highly active photocatalytic ZnO nanocrystalline rods supported on polymer fiber mats: synthesis using atomic layer deposition and hydrothermal crystal growth. Appl Catal A Gen 407:211–216
71. Hong J, He Y (2014) Polyvinylidene fluoride ultrafiltration membrane blended with nano-ZnO particle for photo-catalysis self-cleaning. Desalination 332:67–75
72. Moafi HF, Shojaie AF, Zanjanchi MA (2011) Semiconductor-assisted self-cleaning polymeric fibers based on zinc oxide nanoparticles. J Appl Polym Sci 121:3641–3650
73. Sun L, Shi Y, Li B, Li X, Wang Y (2013) Preparation and characterization of polypyrrole/ TiO_2 nanocomposites by reverse microemulsion polymerization and its photocatalytic activity for the degradation of methyl orange under natural light. Polym Compos 34:1076–1080
74. Ansari MO, Khan MM, Ansari SA, Cho MH (2015) Polythiophene nanocomposites for photodegradation applications: past, present and future. J Saudi Chem Soc 19:494–504
75. Aizawa M, Watanabe S, Shinohara H, Shirakawa H (1985) Photodoping of polyacetylene films. J Chem Soc, Chem Commun 2:62–63
76. Yang Y, Dan Y (2006) Preparation of poly(methyl methacrylate)/titanium oxide composite particles via in-situ emulsion polymerization. J Appl Polym Sci 101:4056–4063
77. Zhang Z, Zheng T, Xu J, Zeng H (2016) Polythiophene/Bi2MoO6: a novel conjugated polymer/nanocrystal hybrid composite for photocatalysis. J Mater Sci 51:3846–3853
78. Wang X, Maeda K, Thomas A, Takanabe K, Xin G, Carlsson JM, Domen K, Antonietti MA (2009) A metal-free polymeric photocatalyst for hydrogen production from water under visible light. Nat Mater 8:76–80
79. Qiua J, Zhanga X, Feng Y, Zhang X, Wang H, Yao J (2018) Modified metal-organic frameworks as photocatalysts. Appl Catal B 231:317–342
80. Ullah H, Tahir AA, Mallick TK (2017) Polypyrrole/TiO_2 com composites for the application of photocatalysis. Sens Actuators B Chem 241:1161–1169
81. Colmenares JC, Kuna E (2017) Photoactive hybrid catalysts based on natural and synthetic polymers: a comparative overview. Molecules 22:790
82. Magalhães F, Mourab FCC, Lago RM (2011) TiO_2/LDPE composites: a new floating photocatalyst for solar degradation of organic contaminants. Desalination 276:266–271
83. Naskar S, Pillay AS, Chanda M (1998) Photocatalytic degradation of organic dyes in aqueous solution with TiO_2 nanoparticles immobilized on foamed polyethylene sheet. J Photochem Photobiol A Chem 3:257–264
84. Jin L, Wu H, Morbidelli M (2015) Synthesis of water-based dispersions of polymer/TiO_2 hybrid nanospheres. Nanomaterials 5:1454–1468

85. Nabid MR, Golbabaee M, Moghaddam AB, Dinarvand R, Sedghi R (2008) Polyaniline/TiO_2 nanocomposite: Enzymatic synthesis and electrochemical properties. Int J Electrochem Sci 3:1117–1126
86. Li H, Fu S, Peng L (2013) Surface modification of cellulose fibers by layer-by-layer self-assembly of lignosulfonates and TiO_2 nanoparticles: effect on photocatalytic abilities and paper properties. Fibers Polym 14:1794–1802
87. Zeng J, Liu S, Cai J, Zhang L (2010) TiO_2 immobilized in cellulose matrix for photocatalytic degradation of phenol under weak UV light irradiation. J Phys Chem C 114:7806–7811
88. Yu DH, Yu X, Wang C, Liu XC, Xing Y (2012) Synthesis of natural cellulose-templated TiO_2/Ag nanosponge composites and photocatalytic properties. ACS Appl Mater Interfaces 4:2781–2788
89. Ahmadizadegan H (2017) Surface modification of TiO_2 nanoparticles with biodegradable nanocellolose and synthesis of novel polyimide/cellulose/TiO_2 membrane. J Colloid Interface Sci 491:390–400
90. Foruzanmehr MR, Vuillaume PY, Robert M, Elkoun S (2015) The effect of grafting a nano-TiO_2 thin film on physical and mechanical properties of cellulosic natural fibers. Mater Des 85:671–678
91. Abdal-hay A, Makhlouf ASH, Khalil KA (2015) Novel, facile, single-step technique of polymer/TiO_2 nanofiber composites membrane for photodegradation of methylene blue. ACS Appl Mater Interfaces 7(24):13329–13341
92. Zhang X, Wang P, Han Q, Li H, Wang T, Ding M (2018) Metal–organic framework based in-syringe solid-phase extraction for the on-site sampling of polycyclic aromatic hydrocarbons from environmental water samples. J Sep Sci 41. https://doi.org/10.1002/jssc.201701383
93. Adeniji AO, Okoh OO, Okoh AI (2017) Petroleum hydrocarbon profiles of water and sediment of Algoa Bay, Eastern Cape, South Africa. Int J Environ Res Public Health 14:1263
94. Martins AF, da Silva DS, Mejía ACC, Bravo JE (2018) Occurrence of polycyclic aromatic hydrocarbons in surface water and hospital wastewater. J Environ Sci Health A. https://doi.org/10.1080/10934529.2017.1422955
95. Bai H, Zhou J, Zhang H, Tang G (2017) Enhanced adsorbability and photocatalytic activity of TiO_2-graphene composite for polycyclic aromatic hydrocarbons removal in aqueous phase. Colloids Surf B 150:68–77
96. Singh RK, Kumar S, Kumar S, Kumar A (2008) Biodegradation kinetic studies for the removal of p-cresol from wastewater using Gliomastix indicus MTCC 3869. Biochem Eng J 40:293
97. Nuhoglu A, Yalcin B (2005) Modelling of phenol removal in a batch reactor. Proc Biochem 40:1233
98. Salim NE, Jaafar J, Ismail AF, Othman MHD, Rahman MA, Yusof N, Qtaishat M, Matsuura T, Aziz F, Salleh WNW (2018) Preparation and characterization of hydrophilic surface modifier macromolecule modified poly (ether sulfone) photocatalytic membrane for phenol removal. Chem Eng J 335:236–247
99. Moslehyani A, Ismail AF, Othman MHD, Matsuura T (2015) Hydrocarbon degradation and separation of bilge water via a novel TiO2-HNTs/PVDFbased photocatalytic membrane reactor (PMR). RSC Adv 5:14147–14155
100. Kümmerer K (2009) Antibiotics in the aquatic environment—a review—part I. Chemosphere 75:417
101. Regitano JB, Leal RMP (2010) Diversity and efficiency of *bradyrhizobium* strains isolated from soil samples collected from around *sesbania virgata* roots using cowpea as trap species. Rev Bras Cienc Solo 34:601
102. Viseras J (1999) Plan de higiene rural. Dossier: problematica de los ´ envases de fitosanitarios. Phytoma Espana 111:12–16

103. Gao D, Liu N, Li W, Han Y (2018) Fabrication of nanoporous polymeric crystalline TiO_2composite for photocatalytic degradation of aqueous organic pollutants under visible light irradiation. Appl Organomet Chem 32. https://doi.org/10.1002/aoc.4119
104. Wang H, Wang N, Wang B, Zhao Q, Fang H, Fu C, Tang C, Jiang F, Zhou Y, Chen Y, Jiang Q (2016) Antibiotics in drinking water in shanghai and their contribution to antibiotic exposure of school children. Environ Sci Technol 50(5):2692–2699
105. Karaoliaa P, Michael-Kordatoua I, Hapeshia E, Drosouc C, Bertakis Y, Christofilosd D, Armatas GS, Sygellou L, Schwartz T, Xekoukoulotakis NP, Fatta-Kassinos D (2018) Removal of antibiotics, antibiotic-resistant bacteria and their associated genes by graphene-based TiO_2 composite photocatalysts under solar radiation in urban wastewaters. Appl Catal B 224:810–824
106. Kong H, Song J, Jang J (2010) Photocatalytic antibacterial capabilities of TiO_2—biocidal polymer nanocomposites synthesized by a surface-initiated photopolymerization. Environ Sci Technol 44(14):5672–5676
107. Cabral JPS (2010) Water microbiology. Bacterial pathogens and water. Int J Environ Res Public Health 7(10):3657–3703
108. Pandey PK, Kass PH, Soupir ML, Biswas S, Singh VP (2014) Contamination of water resources by pathogenic bacteria. AMB Express 4:51
109. Fernández-Ibáñez P, Blanco J, Sichel C, Malato S (2005) Water disinfection by solar photocatalysis using compound parabolic collectors. Catal Today 101:345–352
110. Matsunaga T, Tomoda R, Nakajima T, Wake H (1985) Photoelectrochemical sterilization of microbial cells by semiconductor powders. FEMS Microbiol Lett 29:211–214
111. Saito T, Iwase T, Horie J, Morioka T (1992) Mode of photocatalytic bactericidal action of powdered semiconductor TiO_2 on mutans streptococci. J Photochem Photobiol B Biol 14:369–379
112. Maness P-C, Smolinski S, Blake DM, Huang Z, Wolfrum EJ, Jacoby WA (1999) Bactericidal activity of photocatalytic TiO_2 reaction: toward an understanding of its killing mechanism. Appl Environ Microbiol 65:4094–4098
113. Huang Z, Maness P-C, Blake DM, Wolfrum EJ, Smolinski SL, Jacoby WA (2000) Bactericidal mode of titanium dioxide photocatalysis. J Photochem Photobiol A Chem 130:163–170
114. Benabbou AK, Guillard C, Pigeot-Remy S, Cantau C, Pigot T, Lejeune P (2011) Water disinfection using photosensitizers supported on silica. J Photochem Photobiol A Chem 219:101–108
115. Zan L, Fa W, Peng T, Gong ZK (2007) Photocatalysis effect of nanometer TiO2 and TiO2-coated ceramic plate on Hepatitis B virus. J Photochem Photobiol B Biol 86:165–169
116. Sichel C, de Cara M, Tello J, Blanco J, Fernández-Ibáñez P (2007) Solar photocatalytic disinfection of agricultural pathogenic fungi: Fusarium species. Appl Catal B Environ 74:152–160
117. Sokmen M, Degerli S, Aslan A (2007) Photocatalytic disinfection of Giardia intestinalis and Acanthamoeba castellani cysts in water. Exp Parasitol 119:44–48
118. Szekeres GP, Németh Z, Schrantz K, Nemeth K, Schabikowski M, Traber J, Pronk W, Hernadi K, Graule T (2018) Copper-coated cellulose-based water filters for virus retention. ACS Omega 3(1):446–454
119. Lee M, Chen BY, Den W (2015) Chitosan as a natural polymer for heterogeneous catalysts support: a short review on its applications. Appl Sci 5:1272–1283
120. Luo L, Yang LC, Xiao M, Bian L, Yuan B, Liu Y, Jiang F, Pan X (2015) A novel biotemplated synthesis of TiO2/wood charcoal composites for synergistic removal of bisphenol A by adsorption and photocatalytic degradation. Chem Eng J 262:1275–1283
121. Ohtani N, Tonoi M (2014) Improved photoluminescence lifetime of organic emissive materials embedded in organic-inorganic hybrid thin films fabricated by sol-gel method using tetraethoxysilane. Mol Cryst Liq Cryst 599:132–138

122. Yan SC, Lv SB, Li ZS, Zouabd ZC (2010) Organic-inorganic composite photocatalyst of g-C(3)N(4) and TaON with improved visible light photocatalytic activities. Dalton Trans 39:1488–1491
123. Corma A, Navarro MT, Rey F, Ruiz VR, Sabater MJ (2010) Direct synthesis of a photoactive inorganic-organic mesostructured hybrid material and its application as a photocatalyst. Chem Phys Chem 10:1084–1089
124. Foruzanmehr MR, Vuillaum PY, Elkoun S, Robert M (2016) Physical and mechanical properties of PLA composites reinforced by TiO_2 grafted flax fibers. Mater Des 106: 295–304

Chapter 7
Polymers as Water Disinfectants

Chin Wei Lai, Kian Mun Lee, Bey Fen Leo, Christelle Pau Ping Wong and Soon Weng Chong

Abstract Today, microbial infection appeared as one of the most critical environmental pollutions from our water stream. Indeed, the rising of public awareness for water pollution and water security has urged both researchers and industries to develop cost-effective antimicrobial polymer system. Although a range of polymers have antimicrobial properties, the most frequently studied polymer for water disinfection is chitosan. It offers several advantages, including biodegradable, non-toxic in nature, biocompatible and inexpensive, as compared to other low molecular weight antimicrobial polymers. In general, low molecular weight antimicrobial agents suffer several disadvantages, such as toxicity to the environment and short-term antimicrobial ability. Moreover, using chitosan biopolymer could enhance the efficacy of some existing antimicrobial agents and antifungal agents and minimize the environmental problems. In this chapter, the brief introduction of chitosan as well as modified chitosan on the development of water disinfection is extensively discussed. In particular, this chapter discusses the physicochemical properties of chitosan and different synthesis approaches for chitosan.

7.1 Introduction

Water is the most essential natural resource on earth, and freshwater is necessary for human or other living creatures' survival. However, rapid development of countries leads to the water pollution. Large number of water pollutants such as microbial pathogens, heavy metals, and dyes are dangerous pollutants for human being due to its high toxicity, carcinogenicity, and pathogenicity [1]. The rising of public awareness for water pollution has urged researchers to develop antimicrobial polymer system. Antimicrobial polymer has the ability to inhibit the growth of

C. W. Lai (✉) · K. M. Lee · B. F. Leo · C. P. P. Wong · S. W. Chong
Nanotechnology & Catalysis Research Centre (NANOCAT), University of Malaya (UM), 50603 Kuala Lumpur, Malaysia
e-mail: cwlai@um.edu.my

R. Das (ed.), *Polymeric Materials for Clean Water*,
Springer Series on Polymer and Composite Materials,
https://doi.org/10.1007/978-3-030-00743-0_7

microorganisms such as fungi, bacteria, and virus. Antimicrobial polymers kill bacteria through five steps: (i) Polymer is adsorbed onto the bacterial cell wall; (ii) the antimicrobial polymer diffuses through the cell wall; (iii) antimicrobial polymers bind into the cytoplasmic membrane; (iv) disruption and disintegration of cytoplasmic membrane; and (v) rupture of the cell membrane [2]. Most of the bacteria surfaces are negatively charged; therefore, positively charged antimicrobial polymer is widely used in water disinfection. Chitosan is a natural non-toxic biopolymer prepared by the deacetylation of chitin. Chitosan has attracted great attention due to its antimicrobial and antifungal activity in acidic solution. This is because chitosan is soluble only in acidic solution with pH <6. There are several factors that influence the antimicrobial activity of chitosan, including the type of chitosan, the degree of deacetylation, and the physicochemical properties of chitosan. This chapter is aimed to present an overview of the formation and physicochemical properties of chitosan. The effect of physicochemical properties of chitosan toward the antimicrobial activity in wastewater treatment is also emphasized. In order to further explore the properties of chitosan, a detailed review on modified chitosan in wastewater treatment is presented, including beading, cross-linking, grafting, and surface impregnation.

7.2 Antimicrobial Polymers for Controlling Water Microbes

Numerous antimicrobial polymers have been developed to improve the antimicrobial activity. The ideal antimicrobial polymers should possess the following characteristics: (i) ease and inexpensively synthesized; (ii) wide spectrum of antimicrobial activity; (iii) does not decompose to toxic products; (iv) biodegradable and biocompatible. Different antimicrobial polymers for controlling water microbes have been summarized in Table 7.1. Antimicrobial agents are designed based on polymers that contain antimicrobial functional groups such as quaternary ammonium salt, halogen, alkyl, acrylic. Refer to Table 7.1, poly1-Cl and poly-Br exhibited in killing *Staphylococcus aureus* and *Escherichia coli* cells. However, poly-Br is not suitably utilized in disinfecting potable water due to the emission of toxic bromine [3]. Damm et al. [4] reported that polyamide 6 (PA6) exhibits only a weak antimicrobial efficacy against *E. coli*. Thus, the concentration of bacteria is twofold higher than the initial concentration ($1.8 \pm 0.2 \times 10^6$ CFU/mL) after 24 h of contact time. But, it started to decrease with increasing the weight percentage of silver nanoparticles. In contrast, PA6/silver microcomposite showed poor antimicrobial efficacy even with the highest silver content of 1.9 wt%. This is because of its low specific surface area. In addition, polymers act as matrix of the materials holding the antimicrobial agents [5]. Hence, their chain lengths must be sufficiently long and flexible in order to penetrate into the bacterial cell walls. Tiller et al. [6] investigated the lengths of PVP chains varying from propyl to hexadecyl.

Table 7.1 Antimicrobial activity of antimicrobial polymers

Materials	Concentration of bacteria	Treatment methods	Research highlights	References
Poly[1,3-dichloro-5-methyl-5-(4'-vinylphenyl) hydantoin] (poly1-Cl) and poly [1,3-dibromo-5-methyl-5-(4'-vinylphenyl) hydantoin] (poly1-Br)	• *S. aureus* = 6.9×10^6 CFU/mL • *E. coli* = 8.5×10^6 CFU/mL	Chlorination and bromination	• Poly-Cl displayed a killing efficiency of 6.8 and 6.9 log reduction (99.9999%) against *S. aureus* and *E. coli* in 1.1 s of contact time, respectively. • Poly-Br also showed the same results as Poly-Cl under the same time	[3]
Polyamide 6/silver nanoparticles and microcomposite	• *E. coli* = $1.8 \pm 0.2 \times 10^6$ CFU/mL	N/A	• Polyamide 6/silver nanoparticle with minimum 0.06 wt% of Ag was able to kill 100% of *E. coli*. In contrast, only about 80% of the bacteria was killed by polymer that contained 1.9 wt% of silver microcomposite	[4]
Poly-(N-benzyl-4-vinylpyridinium bromide) (PVP)	• *S. aureus* = 1×10^6 cells/mL • *E. coli* = 1×10^7 cells/mL	Spray	• Hexyl-PVP resulted in >94% and >99% killing efficiency against the *S. aureus* cells and *E. coli* cells, respectively	[6]
Polymeric silsesquioxanes chloride	• *S. aureus* = 4.45×10^5 cells/mL	Membranes	• Polymeric silsesquioxanes chloride (0.33 wt%) has great value in inhibiting the growth of *S. aureus* from 4.45×10^5 to less than 10^1 at 3 days	[7]
ZnPcS/chitosan and p-TAPP/chitosan	• *E. coli* = 1.99×10^5 CFU/mL	Photoreactor	• The ZnPcS/chitosan was able to kill up to 100% of *E. coli* while the p-TAPP/chitosan was only able to kill 99.3% of *E. coli* at 30 min	[8]
Halogenated-polyepicyanuriohydrin and non-halogenated-polyepicyanuriohydrin	• *S. aureus* = 8.9×10^6 CFU/mL • *E. coli* = 9.8×10^6 CFU/mL	Multi-filtration	• Halogenated-polyepicyanuriohydrin was found to result in 9 log reduction (almost 100%) for both *S. aureus* and *E. coli* in 1.5 h, while no antimicrobial activity was observed for non-halogenated polymer	[11]
Poly(glycidyl methacrylate)	• *E. coli* = 6×10^4 cells/mL	Cut plug	• Efficiency activity against *E. coli* that resulted in 99.9% reduction at tributyl phosphonium salt of modified poly(glycidyl methacrylate) concentration of 10 mg/mL was demonstrated	[12]

The results showed that hexyl-PVP exhibited the highest killing efficiency against various airborne bacteria on contact. This is probably related to the visual appearance of the alkylated. Decyl, dodecyl, and hexadecyl-PVP were cloudy, while propyl, butyl, and hexyl-PVP were clear. The cloudiness is probably due to the aggregation of polymer leading to poor interaction with bacterial cells. Polymeric silsesquioxane chloride has high killing efficiency for *S. aureus* cell [7]. The killing efficiency increased with increasing the weight percentage of polymer (0.1, 0.33, and 1.0 wt%). With 0.1 wt% of polymer, it took 7 days to eliminate bacteria. Moreover, silsesquioxanes are insoluble in water as well as in solvent. Hence, it is only suitable in non-aqueous system. Bonnett et al. [8] incorporated the photosensitizers into chitosan membrane for water disinfection systems. The results showed that zinc (II) phthalocyanine tetrasulfonic acid tetrasodium salt (ZnPcS)/chitosan was the most effective in killing the *E. coli* as compared with 5,10,15,20-tetrakis(p-aminophenyl)porphyrin (p-TAPP) photosensitizer.

Among antimicrobial polymers, chitosan is the only natural antimicrobial agent. It has attracted great interest due to its advantages of biocompatible, biodegradable, safe, non-toxic, abundant, and physically and biologically functional characteristics [9]. The antimicrobial activity depends on several factors such as molecular weight, degree of deacetylation, solubility, pH, concentration, and positive charge density. For example, higher degree of deacetylation provides more number of amine groups, thus increasing the efficiency against bacteria. As well known, the amine groups of chitosan are strongly reactive with the functional groups of cell membrane [10].

7.3 Properties of Chitosan

Chitosan, a biopolymer, offers several advantages such as biodegradable, non-toxic, biocompatible, and inexpensive [13]. Each glucosamine unit contains a free amino group and hydroxyl group, which give amazing properties to the chitosan as shown in Fig. 7.1.

Chitosan is widely used in various industries, including pharmaceuticals, food industries, agricultural, and water filtration. However, there are several factors that affect the physicochemical properties of chitosan as shown in Fig. 7.2.

(a) **Degree of deacetylation (DD %):**

DD is the total number of N-deacetylated sites that are present on biopolymer's backbone. This number can be determined using Fourier transform infrared (FTIR) spectroscopy [16]. The percentage of DD depends on the reaction time, reaction temperature, concentration of alkali agent, the use of successive baths, atmospheric conditions, and the alkali and reducing agent in deacetylation reaction [17]. Among the aforementioned parameters, the use of successive baths, reaction time, and concentration of alkali agent gives significant effect to the DD %. The DD % affects

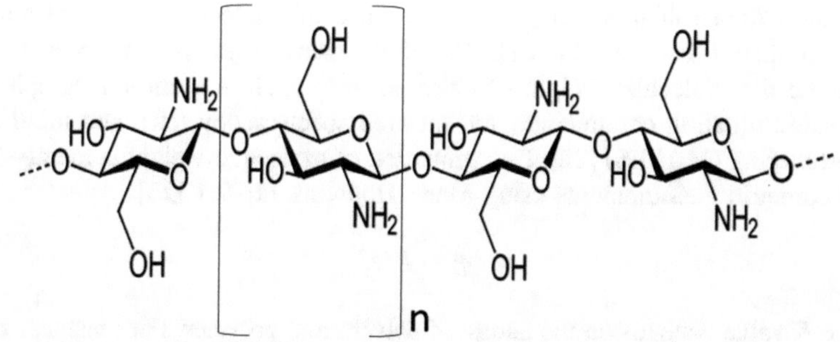

Fig. 7.1 Structure of chitosan [14]

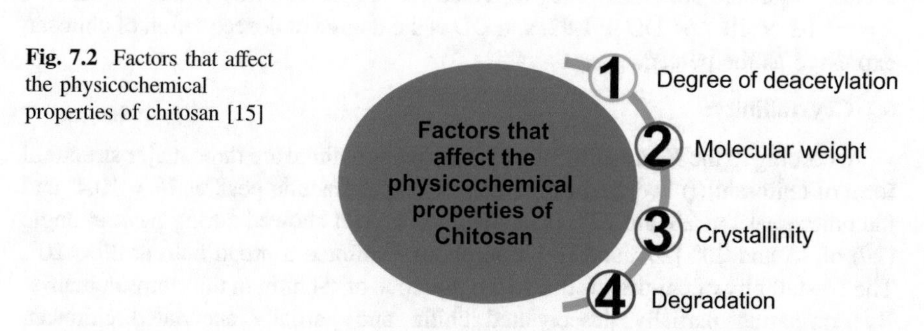

Fig. 7.2 Factors that affect the physicochemical properties of chitosan [15]

the presence of amino groups, solubility, crystallinity, and molecular weight as well as the capability of adsorption in removing pollutants from wastewater [18].

Solubility of chitosan is dependent on the DD %. Chitosan is insoluble in water and alkaline solutions, owing to the amino groups of the biopolymer participating in intra- as well as inter-molecular of hydrogen bond. Chitosan is easily dissolved in the acidic solution with $pH < 6$ and certain organic solvents (i.e., dimethyl sulfoxide and dimethyl formamide) due to the protonation of its amino groups, making it positively charged and become water-soluble cationic polyelectrolyte [19]. However, the amino group of chitosan deprotonated at $pH > 6$, thus making it insoluble. The soluble–insoluble transition occurs at its logarithmic acid dissociation constant (pKa) around pH between 6 and 6.5 [20].

(b) **Molecular weight:**

The molecular weight of chitosan is also greatly influenced by the DD %. The average molecular weight of chitin is 1.03×10^6 to 2.5×10^6 Da. After the deacetylation reaction, the molecular weight of chitin will decrease from 1×10^5 to 5×10^5 Da. In other words, transforming chitin into chitosan decreases the molecular weight. Researchers found that chitosan has molecular weight in ranges of 5×10^4 to 2×10^5 Da. Its molecular weight plays an important role in determining the antimicrobial properties, such as permeability through the cell wall. The antimicrobial property increased with molecular weight lower than 5×10^4 Da, while the antimicrobial property decreased when the molecular weight of polymer is higher than 1.2×10^5 Da [21]. Therefore, many methods have been used to determine the molecular weight of chitosan: size exclusion chromatography, gel permeation/filtration chromatography, infrared spectroscopy (IR), and multi-angle light scattering (MALLS) [22]. Determination of molecular weight is mainly based on viscometric measurements using Mark–Houwink Eq. 7.1 [23]:

$$[\eta] = KM_v^a \quad (7.1)$$

The K value depends on the nature of solvent and polymer. For instance, Zhang and Neau [24] prepared chitosan solution in 0.2 M acetic acid/0.1 M sodium acetate aqueous solution. The K value is equal to $1.64 \times 10^{-30} \times DD^{14}$, $a = -1.02 \times 10^{-2} \times DD + 1.82$, and DD is the degree of deacetylation of chitosan expressed as the percentages.

(c) **Crystallinity:**

According to the X-ray diffraction (XRD) pattern, there are three major structural form of chitosan: (i) hydrated exhibited a strong diffraction peak at $2\theta = 10.4°$ and the other peaks at 20 and 22°; (ii) anhydrous crystal showed strong peak at angle (2θ) of 15 and 20° [25]; and (iii) amorphous exhibited a broad halo at $2\theta = 20°$. The crystallinity of chitosan is due to the presence of α-chitin in the chitosan matrix. By comparing partially deacetylated chitin and partially acetylated chitosan, researchers had concluded that chitosan with 100% of DD is purely crystalline, but any N-acetylation present in the polymer may lead to the decrease in crystallinity of the polymer [26]. Chitosan in semicrystalline form is widely used in various applications such as medical and wastewater treatment. Chitosan can be synthesized using three methods, namely chemical deacetylation, enzymatic degradation, and physical degradation methods [27]. This chemical, biochemical, and physical methods result in changes of physicochemical properties of chitosan. Further explanation of degradation methods is presented in Sect. 7.5.

7.4 Chitosan Synthesis

Chitin, the main source of chitosan, was first discovered and isolated from mushrooms in 1811, but the name of "chitin" shows up when it was extracted from insect in the 1830s. Chitin is the second-most abundant natural biopolymer after cellulose as it occurs as a component in exoskeleton of arthropods (i.e., crabs and shrimps), fungi cell walls, and plankton [28]. Chitin was first prepared by hydrolyzing the surface flesh of shrimp waste using 0.5 M NaOH at ambient temperature (Fig. 7.3). The alkali-treated waste was then washed, dried, and grounded to obtain powder that can pass through 250 μm sieve. After that, the powder was subjected to demineralization and deproteinization. During the demineralization process, the powder was soaked in 0.25 M HCl solution at ambient temperature until the CO_2

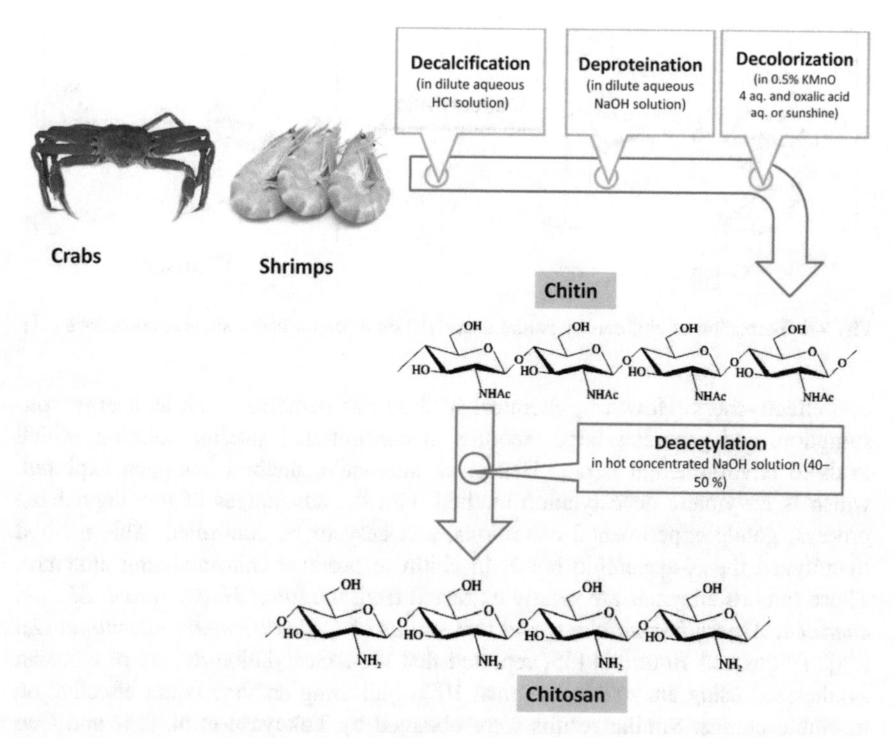

Fig. 7.3 Production method for chitin and chitosan

gas was discharged. Several acids (e.g., HCl, HNO_3, H_2SO_4, CH_3COOH, and HCOOH) have been tested as demineralization reagents in removing of calcium carbonate [29]. The duration and CO_2 emission were dependent on the species. The demineralization powder was treated in 1 M NaOH at 70 °C to hydrolyze the protein. A wide range of chemicals have been performed as deproteinization reagents including NaOH, Na_2CO_3, $NaHCO_3$, Na_2SO_3, $NaHSO_3$, Na_2S, Na_3PO_4, KOH, K_2CO_3, $Ca(OH)_2$, and $CaHSO_3$. Among these reagents, NaOH is preferable. The absence of proteins was indicated by the absence of color of the medium [30].

Chitin is a long-chain polymer of an N-acetyl-D-glucosamine that is linked together by β-(1→4) linkage, and the most famous deacetylated derivative of chitin is chitosan with $-NH_2$ that occurs on C-2 position of D-glucosamine repeat unit as illustrated in Fig. 7.4.

In chemical deacetylation method, it can be classified into two types: homogeneous and heterogeneous methods. In homogenous method, 3 g of chitin is first dispersed in concentrated solution of NaOH in ratio of 30 g NaOH: 45 g H_2O at 25 °C for 3 h or more followed by dissolution in crushed ice at 0 °C. This method produced chitosan with an average degree of deacetylation of 48–55% [32], while the degree of deacetylation of chitosan that was prepared through heterogeneous method can be up to 85–99% [33]. Chemical deacetylation methods are widely used in the preparation of chitosan, owing to its large-scale production and

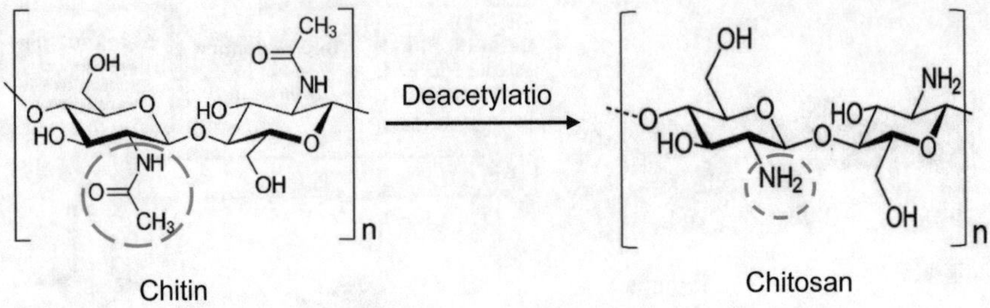

Fig. 7.4 Formation of chitosan by partial deacetylation of chitin under alkaline conditions [31]

cost-effectiveness. However, chemical method has demerits, such as energy consumption, and generates large amounts of concentrated alkaline solution, which leads to environmental issues. Hence, an alternative method has been explored, which is enzymatic deacetylation method with the advantages of non-degradable process, gentle experimental conditions, and easy to be controlled. This method hydrolyzed the N-acetamido bonds in chitin to produce chitosan-using enzymes. Those famous enzymes are mostly extracted from the fungi *Mucor rouxii, Absidia coerulea, Aspergillus nidulans,* and two strains of *Colletotrichum lindemuthianum* [34]. Tsigos and Bouriotis [35] reported that the deacetylation degree of chitosan synthesized using enzyme is less than 10%, indicating enzyme is not effective on insoluble chitins. Similar results were obtained by Tokuyasu et al. [34] and Gao et al. [36] using chitin deacetylases isolated from other sources. Moreover, the high cost of enzymes inhibits their use in industrial production of chitosan. Furthermore, chitosan also can be prepared using physical methods (i.e., oxidation, hydrolytic, thermal, and radiation degradation). Physical method is considered as an energy-saving, environment-friendly, and effective method, because it requires low energy to break the chemical bond. Muzzarelli and Rocchetti [37] claimed that sonication over a long period of treatment leads to immediate chain degradation and to detectable deacetylation.

7.5 Chitosan in Water Disinfection

Water can be a medium of disease transmission, for instance, cholera, diarrhea, typhus. Inadequate management of agricultural, industrial, and urban wastewater leads to the contamination of drinking water due to the presence of bacteria (e.g., *E. coli*), viruses, fungi, or parasites. Thus, water disinfection must be performed to eliminate these microorganisms by physical or chemical means using chemical substances such as gaseous chlorine, calcium hypochlorite, chlorinated lime, and sodium hypochlorite. Although the hypochlorous acid and hypochlorite ion formed from gaseous chlorine or sodium hypochlorite have a biocidal effect, chlorinated compounds can be harmful to human if the concentration exceeds the safe chlorine level standard defined by the World Health Organization (WHO). According to WHO 2014's report on "Progress on Sanitation and Drinking Water," approximately 748 million and 2.5 billion people do not have access to clean drinking water and sanitized water, respectively [38]. Thus, the lack of water disinfection is rendering individuals more vulnerable to diarrheal disease. Ultimately, it is essential to develop inexpensive and easy water treatment solutions to reduce global diarrheal mortality [39]. The increasing demand for clean water requires better treatments, lower operating cost, high reliability, and high energy efficiency. In this respect, chitosan, a natural biopolymer, has emerged as an effective alternative for water disinfection. Monica et al. reported a new approach for the inactivation of microorganisms which utilized cotton gauzes coated with chitosan as a water filter for biological disinfection against Gram-negative and Gram-positive bacteria [40].

Chitosan is a natural cellulosic material that can be involved in radical grafting, which is also easily available and of low production cost. Besides that, chitosan has shown the impressive performance of inhibition of *S. aureus* and *K. pneumoniae* with 4 and 8 s reaction times, respectively. Chen et al. proved that chitosan is effective in eliminating waterborne pathogens, especially for Gram-negative bacteria. A higher degree of deacetylation of chitosan has shown better bacteria inhibition effectiveness. The author also found that the antibacterial activity of chitosan depends on their molecular weight. Various growth phases of *E. coli* were identified, and it was found that the negative surface charges will interact with chitosan, destroying the bacteria itself. Up-to-date, literature has suggested that chitosan promotes chemical coagulation, floc formation, and sedimentation of viruses and colloidal particles in the contaminated water via interparticle bridging and charge neutralization [28, 41]. The coagulation effectiveness of chitosan is determined by parameters such as the degree of deacetylation, molecular weight, surface charge, and pH. The amino group on chitosan is identified to be responsible for the adsorption of negatively charged colloids, for instance, clay turbidity, bacteria, and viruses.

Wastewater is produced by a vast spectrum of domestic and industrial activities containing various organic, inorganic, and biological contaminants. These contaminants cause health hazards without the proper treatment. Research in wastewater treatment has been emphasizing on easy accessibility and environmentally sustainable alternatives to remove toxic and persistent chemicals from industrial effluents. Conventional water treatment system consists of the following processes: coagulation, flocculation, sedimentation, filtration, and disinfection [42]. Figure 7.5 shows the overall process of the application of chitosan in water treatment.

Coagulation is the first step to remove microorganisms, colloidal natural organic matter (NOM), turbidity, and metals [43–45]. In the coagulation process, electronegative colloidal particles such as dyes, microorganisms, clays, and NOM are destabilized and agglomerated to form flocs that will precipitate and sediment. The addition of coagulant causes reduction of the repulsive electrical potential of the electronegative colloidal particles [46]. Colloids are removed through a chain of processes: charge neutralization, adsorption, formation of metal complexes and

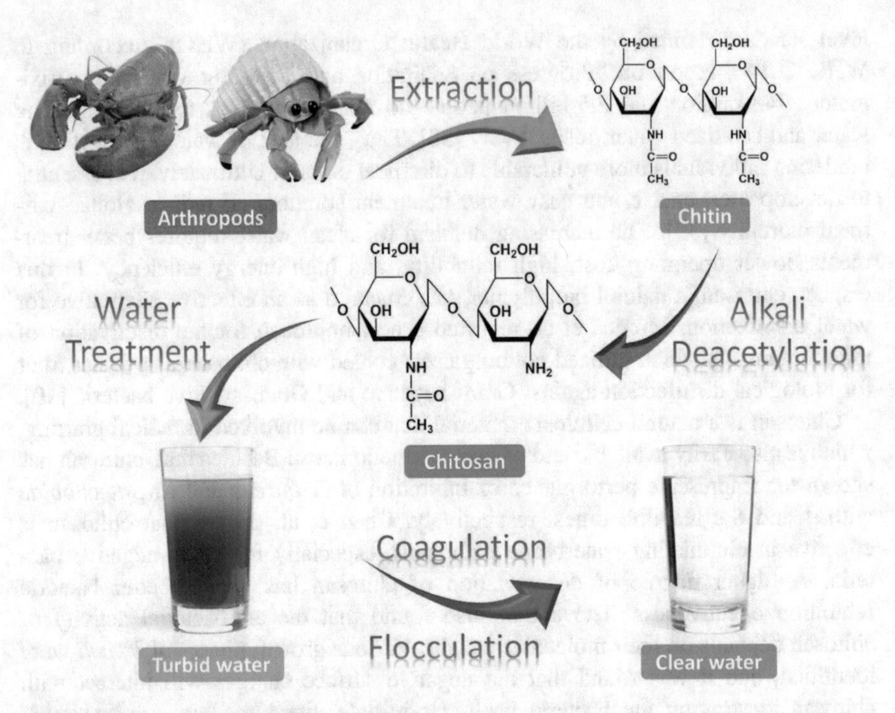

Fig. 7.5 Water treatment process using chitosan: coagulation, flocculation, sedimentation, filtration, and disinfection

precipitation [45]. The use of inorganic coagulant salts (e.g., aluminum chloride and ferric chloride) shows some drawbacks such as high levels of residual that might cause neurological diseases and toxic inorganic waste. Thus, proper disposal or addition of chemicals is required to stabilize and control corrosion within the distribution system [45]. Thus, these disadvantages of inorganic coagulant have increased the interest of researchers toward chitosan. Chitosan biopolymers are the world's second-most abundant polysaccharide, produced through the deacetylation of chitin [28, 41]. Chitin is a natural source of biomass which can be found in the exoskeleton of arthropods [41]. Thus, chitosan is a highly potential natural coagulant which is easily available, non-toxic, and biodegradable. Chitosan contains the amino groups making it positively charged, which is a critical factor for efficient removal of negatively charged colloids.

7.5.1 Chitosan Nanocomposite as Superior Water Disinfectant

The advancements in nanotechnology are a promising way to solve the water quality issues by introducing nanomaterials in removing water contaminants, particularly

E. coli, *Pseudomonas aeruginosa*, *Salmonella enteritidis,* and *S. aureus*. It has been reported that *S. enteritidis and S. aureus* can cause diseases such as food poisoning, cellulitis, toxic shock syndrome, and typhoid [47], while *E. coli could* colonize the gastrointestinal tract [48]. Also, *P. aeruginosa* was commonly found in hospital water sources and could lead to death [49]. Hence, the removal of such bacteria from water bodies is inevitable. It is well known that nanomaterials have extraordinary properties such as large surface area, various morphologies, and enhanced catalytic activity for water treatment [50, 51]. However, the use of nanomaterials in industrial applications is limited due to the difficulty in controlling their particle sizes and agglomeration problem. In order to overcome these limitations, scientists have introduced nanoparticles on substrates to ensure a better dispersion of the nanoparticles [52, 53]. Polymer substrates serve as an excellent support for the nanoparticles by controlling the release rate of nanoparticles into the aqueous media, as shown in Fig. 7.6 [54]. Among the polymer substrates, chitosan has been widely used due to its high abundance, cheapness, non-toxicity, and antimicrobial activity [55–57]. Ag nanoparticles are well-known disinfectant that has very strong antimicrobial activity toward Gram-positive or Gram-negative bacteria [58, 59]. However, the difficulty in their storage and aggregation has limited their practical use [60]. Incorporating Ag nanoparticles (NPs) into the chitosan not only overcomes the above problems but also leads to a better antibacterial activity [61]. Lee et al. [62] synthesized Ag NPs/chitosan nanofibers via electrospinning, by varying the amount of Ag NPs. In their study, *P. aeruginosa* and *S. aureus* were selected as the model bacteria. It was found that the composite materials showed higher antimicrobial activity compared to that of chitosan alone. The loading of Ag NPs not only affects the diameter of the nanofibers but also influences the bacterial growth. The bacteria inhibiting effect increases with the quantity of Ag NPs. An et al. [63] prepared Ag/

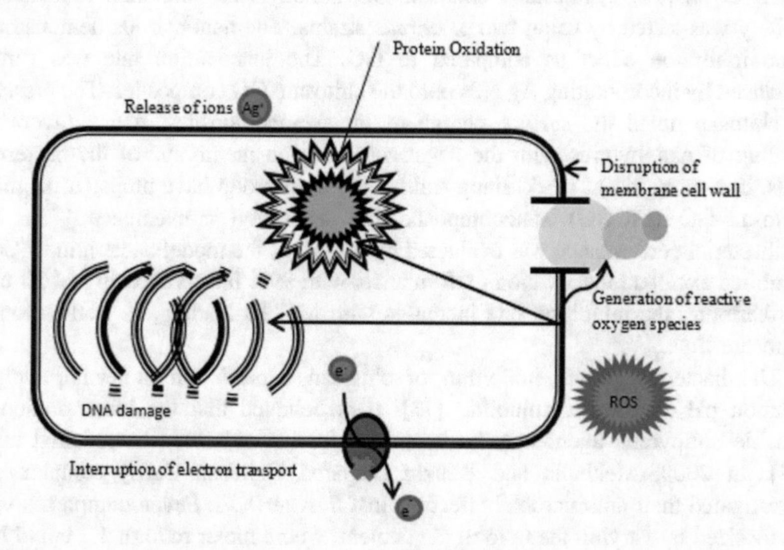

Fig. 7.6 Schematic diagram of mechanism of antimicrobial activities exerted by nanomaterials

chitosan microspheres using an inverse-emulsification cross-linking method, and their antibacterial activity was evaluated by using *E. coli* and *S. aureus*. Experimental results revealed that the As-synthesized microspheres exhibited enhanced inhibitory effect with low Ag contents (2.2 or 4.5 wt%), owing to their smaller size and higher surface area. Punitha and co-researchers have synthesized Ag–chitosan nanocomposites via a simple and eco-friendly method by using β-cyclodextrin as a capping agent [64]. As a result, the Ag–chitosan nanocomposites with spherical shape and less than 1 nm were obtained. The nanocomposites showed effective disinfection against *E. coli* and *S. aureus*, due to the binding of Ag NPs and the Ag^+ with sulfur- and phosphorus-containing biomolecules, which cause cell damage [65]. Kumar-Krishnan et al. [66] prepared Ag–chitosan nanocomposite films via an in situ chemical reduction method. They found that 1 wt% of Ag NPs showed pronounced antibacterial activity toward *E. coli* and *S. aureus*. Higher loading of Ag is not beneficial for bacterial removal due to the reduction in the surface area of NPs. Yang et al. [67] successfully prepared uniform Ag NPs–chitosan composite via one-step microfluidic process. It was found that the composite possesses higher antibacterial activity against *E. coli*. Moreover, the antimicrobial activity is proportional to the amount of Ag NPs loaded, where *E. coli* viability dropped from 275 to 10% on day 5 with 15 mM Ag.

Graphene oxide (GO), another interesting material that has been investigated recently, showed antibacterial activity [68]. Mazaheri et al. [69] fabricated chitosan–GO nanocomposites for the disinfection of *S. aureus*. The experimental results revealed that the composite could suppress the growth of the bacteria up to 77% in only 3 h. Also, the inhibition rate increases with the GO content, which is due to both membrane and oxidative stresses and/or the direct contact of the sharp edges of the GO sheets (vertically aligned ones) with bacterial cell membrane [70]. Marta et al. [71] synthesized chitosan–GO nanohybrids, and their antibacterial activity was tested by using two *S. aureus* strains. The nanohybrids demonstrated good inhibition effect as compared to GO. The inactivation rate was further enhanced by incorporating Ag NPs onto the chitosan–GO composites. The presence of chitosan tuned the surface charge of the nanocomposites, which favors the binding of nanohybrids with the negatively charged membrane of the bacterium cells. In a more recent work, Jiang and his research group have prepared magnetic chitosan–GO (MCGO) nanocomposite for separation convenience [72]. The antibacterial performance was evaluated by *E. coli* as the model bacterium. MCGO exhibited excellent inactivation performance, with 99% inactivation in just 20 min. Furthermore, the inhibition rate increases with MCGO loading, as well as longer exposure time.

The bacterial growth inhibition of chitosan depends on molecular weight, solution pH, and water solubility [73]. It is believed that the incorporation of suitable antipyretic agent may be beneficial for the enhanced bactericidal effect [74]. In 2009, Mekhalia and Bouzid prepared Chitosan–Cu(II) complex and investigated their antimicrobial effect against *S. enteritidis*. Seven complexes were synthesized by varying the ratio of Cu content, where molar ratio of 1:1 could lead to 100% inhibition rate [75]. Ancona et al. [76] synthesized Cu NPs–chitosan hybrid via a femtosecond laser ablation technique. Ultrafine and monodisperse Cu NPs–chitosan hybrid were obtained with chitosan loading of 1 g/L. The nanohybrid also displayed superb inhibition effect against *E. coli* even at only 4 h contact. Arjunan et al. [77] synthesized Cu–chitosan nanocomposite without any chemical reducing agent. The antimicrobial test was conducted on *S. aureus* and *P. aeruginosa*. The experimental results revealed that the nanocomposite was able to inhibit the growth of both microorganisms, with maximum inhibition rate found at 100 μg concentration. The higher surface area and greater adsorption ability of the nanocomposite have contributed to the cell damage. Moreover, the generation of oxy-radicals may also lead to the bacterial death [78].

7.6 Summary

One of the crucial steps taken by many developed countries to aid the earth and minimize the water disinfection issue is applying sustainable development. It seems that the use of chemically modified chitosan or metal-complexed chitosan composites is a strategic goal to solve our current water pollution issue. Indeed, chitosan has garnered great interest from many researchers and industries because of the non-toxicity in nature, biocompatibility, and biodegradability. In this case, these chemically modified chitosan or metal-complexed chitosan composites can be potentially implemented as a tertiary method of treatment in processes like polymer-assisted ultrafiltration to solve our current water disinfection issue. As a matter of fact, chitosan possesses many reactive hydroxyl and primary amine groups, and hence, most of the chemical modifications as well as metal complexes modifications are made with respect to these functional groups of the biopolymer. It can be modified by chemical or metal-complexed substituents that have functional groups in a process known as "grafting." In addition to that, further studies for the regeneration and reuse of chemically modified chitosan or metal-complexed chitosan composites in the water disinfection field are required in order to minimize the costs of the process and improve its chances of being an economically viable option for wastewater treatment and water disinfection applications.

References and Future Readings

1. Wang B, Zhu Y, Bai Z, Luque R, Xuan J (2017) Functionalized chitosan biosorbents with ultra-high performance, mechanical strength and tunable selectivity for heavy metals in wastewater treatment. Chem Eng J 325:350–359
2. Ikeda T, Yamaguchi H, Tazuke S (1984) New polymeric biocides: synthesis and antibacterial activities of polycations with pendant biguanide groups. Antimicrob Agents Chemother 26:139–144
3. Chen Y, Worley S, Kim J, Wei C-I, Chen T-Y, Santiago J, Williams J, Sun G (2003) Biocidal poly (styrenehydantoin) beads for disinfection of water. Ind Eng Chem Res 42:280–284

4. Damm C, Münstedt H, Rösch A (2008) The antimicrobial efficacy of polyamide 6/silver-nano-and microcomposites. Mater Chem Phys 108:61–66
5. Muñoz-Bonilla A, Fernández-García M (2012) Polymeric materials with antimicrobial activity. Prog Polym Sci 37:281–339
6. Tiller JC, Liao C-J, Lewis K, Klibanov AM (2001) Designing surfaces that kill bacteria on contact. Proc Natl Acad Sci 98:5981–5985
7. Isquith AJ, Gettings RL (1986) Insoluble polymeric contact preservatives. Google Patents 1986
8. Bonnett R, Krysteva MA, Lalov IG, Artarsky SV (2006) Water disinfection using photosensitizers immobilized on chitosan. Water Res 40:1269–1275
9. Badawy ME, Rabea EI (2011) A biopolymer chitosan and its derivatives as promising antimicrobial agents against plant pathogens and their applications in crop protection. Int J Carbohydr Chem
10. Ahmed S, Ikram S (2017) Chitosan: derivatives, composites and applications. Wiley
11. Ahmed AE-SI, Cavalli G, Bushell ME, Wardell JN, Pedley S, Charles K, Hay JN (2011) New approach to produce water free of bacteria, viruses, and halogens in a recyclable system. Appl Environ Microbiol 77:847–853
12. Kenawy E-R, Abdel-Hay FI, El-Shanshoury AE-RR, El-Newehy MH (1998) Biologically active polymers: synthesis and antimicrobial activity of modified glycidyl methacrylate polymers having a quaternary ammonium and phosphonium groups. J Controlled Release 50:145–152
13. Peng H, Wu J, Wang Y, Wang H, Liu Z, Shi Y, Guo X (2016) A facile approach for preparation of underwater superoleophobicity cellulose/chitosan composite aerogel for oil/water separation. Appl Phys A: Mater Sci Process 122
14. Wang X, Uchiyama S (2013) Polymers for biosensors construction, in: state of the art in biosensors-general aspects. InTech
15. Ahmed S, Ikram S (2015) Chitosan & its derivatives: a review in recent innovations. Int J Pharm Sci Res 6:14
16. Baxter A, Dillon M, Taylor KA, Roberts GA (1992) Improved method for ir determination of the degree of N-acetylation of chitosan. Int J Biol Macromol 14:166–169
17. Younes I, Ghorbel-Bellaaj O, Chaabouni M, Rinaudo M, Souard F, Vanhaverbeke C, Jellouli K, Nasri M (2014) Use of a fractional factorial design to study the effects of experimental factors on the chitin deacetylation. Int J Biol Macromol 70:385–390
18. Vakili M, Rafatullah M, Salamatinia B, Abdullah AZ, Ibrahim MH, Tan KB, Gholami Z, Amouzgar P (2014) Application of chitosan and its derivatives as adsorbents for dye removal from water and wastewater: a review. Carbohyd Polym 113:115–130
19. Bina B, Mehdinejad M, Nikaeen M, Attar HM (2009) Effectiveness of chitosan as natural coagulant aid in treating turbid waters. J Environ Health Sci Eng 6:247–252
20. Yep T (2016) Application of chitosan in the treatment of wastewater from agricultural sources. University of Windsor (Canada)
21. Kenawy E-R, Worley S, Broughton R (2007) The chemistry and applications of antimicrobial polymers: a state-of-the-art review. Biomacromol 8:1359–1384
22. Tsaih ML, Chen RH (1999) Molecular weight determination of 83% degree of decetylation chitosan with non-Gaussian and wide range distribution by high-performance size exclusion chromatography and capillary viscometry. J Appl Polym Sci 71:1905–1913
23. Wang W, Bo S, Li S, Qin W (1991) Determination of the Mark-Houwink equation for chitosans with different degrees of deacetylation. Int J Biol Macromol 13:281–285
24. Zhang H, Neau SH (2001) In vitro degradation of chitosan by a commercial enzyme preparation: effect of molecular weight and degree of deacetylation. Biomaterials 22:1653–1658
25. Ogawa K (1991) Effect of heating an aqueous suspension of chitosan on the crystallinity and polymorphs. Agric Biol Chem 55:2375–2379
26. Ogawa K, Yui T (1993) Crystallinity of partially N-acetylated chitosans. Biosci Biotechnol Biochem 57:1466–1469

27. Wiśniewska-Wrona M, Niekraszewicz A, Ciechańska D, Pospieszny H, Orlikowski LB (2007) Biological properties of chitosan degradation products. Pol Chitin Soc Monogr 12:149–156
28. Rinaudo M (2006) Chitin and chitosan: properties and applications. Prog Polym Sci 31:603–632
29. Percot A, Viton C, Domard A (2003) Characterization of shrimp shell deproteinization. Biomacromol 4:1380–1385
30. Al Sagheer F, Al-Sughayer M, Muslim S, Elsabee MZ (2009) Extraction and characterization of chitin and chitosan from marine sources in Arabian Gulf. Carbohyd Polym 77:410–419
31. Zargar V, Asghari M, Dashti A (2015) A review on chitin and chitosan polymers: structure, chemistry, solubility, derivatives, and applications. ChemBioEng Rev 2:204–226
32. Chang KLB, Tsai G, Lee J, Fu W-R (1997) Heterogeneous N-deacetylation of chitin in alkaline solution. Carbohyd Res 303:327–332
33. Younes I, Rinaudo M (2015) Chitin and chitosan preparation from marine sources. Struct Prop Appl Mar Drugs 13:1133–1174
34. Tokuyasu K, Ohnishi-Kameyama M, Hayashi K (1996) Purification and characterization of extracellular chitin deacetylase from *Colletotrichum lindemuthianum*. Biosci Biotechnol Biochem 60:1598–1603
35. Tsigos I, Bouriotis V (1995) Purification and characterization of chitin deacetylase from *Colletotrichum lindemuthianum*. J Biol Chem 270:26286–26291
36. Gao X-D, Katsumoto T, Onodera K (1995) Purification and characterization of chitin deacetylase from *Absidia coerulea*. J Biochem 117:257–263
37. Muzzarelli RA, Rocchetti R (1985) Determination of the degree of acetylation of chitosans by first derivative ultraviolet spectrophotometry. Carbohyd Polym 5:461–472
38. W.U.J.W. Supply, S.M. Programme (2014) Progress on drinking water and sanitation: 2014 update. World Health Organization
39. Sobsey MD, Stauber CE, Casanova LM, Brown JM, Elliott MA (2008) Point of use household drinking water filtration: a practical, effective solution for providing sustained access to safe drinking water in the developing world. Environ Sci Technol 42:4261–4267
40. Periolatto M, Ferrero F, Vineis C, Varesano A (2014) Antibacterial water filtration by cationized or chitosan coated cotton gauze. Chem Eng Trans 38:235–240
41. Kumar MNR (2000) A review of chitin and chitosan applications. React Funct Polym 46:1–27
42. Crittenden JC, Trussell RR, Hand DW, Howe KJ, Tchobanoglous G (2012) MWH's water treatment: principles and design. Wiley
43. Budd GC, Hess AF, Shorney-Darby H, Neemann JJ, Spencer CM, Bellamy JD, Hargette PH (2004) Coagulation applications for new treatment goals. J (Am Water Works Assoc) 96:102–113
44. Cheng Y-L, Wong R-J, Lin JC-T, Huang C, Lee D-J, Lai J-Y (2010) Pre-treatment of natural organic matters containing raw water using coagulation. Sep Sci Technol 45:911–919
45. Matilainen A, Vepsäläinen M, Sillanpää M (2010) Natural organic matter removal by coagulation during drinking water treatment: a review. Adv Coll Interface Sci 159:189–197
46. Faust SD, Aly OM (1998) Chemistry of water treatment. CRC Press
47. Teh SJ, Yeoh SL, Lee KM, Lai CW, Abdul Hamid SB, Thong KL (2016) Effect of reduced graphene oxide-hybridized ZnO thin films on the photoinactivation of *Staphylococcus aureus* and *Salmonella enterica* serovar Typhi. J Photochem Photobiol, B 161:25–33
48. Salvadori M, Coleman B, Louie M, McEwen S, McGeer A (2004) Consumption of antimicrobial-resistant *Escherichia coli*-contaminated well water: human health impact. PSI Clin Res:6–25
49. Quick J, Cumley N, Wearn CM, Niebel M, Constantinidou C, Thomas CM, Pallen MJ, Moiemen NS, Bamford A, Oppenheim B (2014) Seeking the source of *Pseudomonas aeruginosa* infections in a recently opened hospital: an observational study using whole-genome sequencing. BMJ Open 4:e006278

50. Raghupathi KR, Koodali RT, Manna AC (2011) Size-dependent bacterial growth inhibition and mechanism of antibacterial activity of zinc oxide nanoparticles. Langmuir 27:4020–4028
51. Warheit DB (2008) How meaningful are the results of nanotoxicity studies in the absence of adequate material characterization? Toxicol Sci 101:183–185
52. Guibal E, Cambe S, Bayle S, Taulemesse J-M, Vincent T (2013) Silver/chitosan/cellulose fibers foam composites: from synthesis to antibacterial properties. J Colloid Interface Sci 393:411–420
53. Motshekga SC, Ray SS (2017) Highly efficient inactivation of bacteria found in drinking water using chitosan-bentonite composites: modelling and breakthrough curve analysis. Water Res 111:213–223
54. Li Q, Mahendra S, Lyon DY, Brunet L, Liga MV, Li D, Alvarez PJ (2008) Antimicrobial nanomaterials for water disinfection and microbial control: potential applications and implications. Water Res 42:4591–4602
55. Chien R-C, Yen M-T, Mau J-L (2016) Antimicrobial and antitumor activities of chitosan from shiitake stipes, compared to commercial chitosan from crab shells. Carbohyd Polym 138:259–264
56. Jana S, Laha B, Maiti S (2015) Boswellia gum resin/chitosan polymer composites: controlled delivery vehicles for aceclofenac. Int J Biol Macromol 77:303–306
57. Van Hoa N, Khong TT, Quyen TTH, Trung TS (2016) One-step facile synthesis of mesoporous graphene/Fe_3O_4/chitosan nanocomposite and its adsorption capacity for a textile dye. J Water Process Eng 9:170–178
58. Ishihara M, Nguyen VQ, Mori Y, Nakamura S, Hattori H (2015) Adsorption of silver nanoparticles onto different surface structures of chitin/chitosan and correlations with antimicrobial activities. Int J Mol Sci 16:13973–13988
59. Ito K, Saito A, Fujie T, Miyazaki H, Kinoshita M, Saitoh D, Ohtsubo S, Takeoka S (2016) Development of a ubiquitously transferrable silver-nanoparticle-loaded polymer nanosheet as an antimicrobial coating. J Biomed Mater Res B Appl Biomater 104:585–593
60. Grumezescu AM, Andronescu E, Holban AM, Ficai A, Ficai D, Voicu G, Grumezescu V, Balaure PC, Chifiriuc CM (2013) Water dispersible cross-linked magnetic chitosan beads for increasing the antimicrobial efficiency of aminoglycoside antibiotics. Int J Pharm 454:233–240
61. Zain NM, Stapley A, Shama G (2014) Green synthesis of silver and copper nanoparticles using Ascorbic acid and Chitosan for antimicrobial applications. Carbohyd Polym 112:195–202
62. Lee SJ, Heo DN, Moon J-H, Ko W-K, Lee JB, Bae MS, Park SW, Kim JE, Lee DH, Kim E-C (2014) Electrospun chitosan nanofibers with controlled levels of silver nanoparticles. Preparation, characterization and antibacterial activity. Carbohyd Polym 111:530–537
63. An J, Ji Z, Wang D, Luo Q, Li X (2014) Preparation and characterization of uniform-sized chitosan/silver microspheres with antibacterial activities. Mater Sci Eng, C 36:33–41
64. Punitha N, Ramesh P, Geetha D (2015) Spectral, morphological and antibacterial studies of β-cyclodextrin stabilized silver–Chitosan nanocomposites. Spectrochim Acta Part A Mol Biomol Spectrosc 136:1710–1717
65. Ahamed M, AlSalhi MS, Siddiqui M (2010) Silver nanoparticle applications and human health. Clin Chim Acta 411:1841–1848
66. Kumar-Krishnan S, Prokhorov E, Hernández-Iturriaga M, Mota-Morales JD, Vázquez-Lepe M, Kovalenko Y, Sanchez IC, Luna-Bárcenas G (2015) Chitosan/silver nanocomposites: synergistic antibacterial action of silver nanoparticles and silver ions. Eur Polymer J 67:242–251
67. Yang C-H, Wang L-S, Chen S-Y, Huang M-C, Li Y-H, Lin Y-C, Chen P-F, Shaw J-F, Huang K-S (2016) Microfluidic assisted synthesis of silver nanoparticle–chitosan composite microparticles for antibacterial applications. Int J Pharm 510:493–500
68. Akhavan O, Ghaderi E (2010) Toxicity of graphene and graphene oxide nanowalls against bacteria. ACS Nano 4:5731–5736

69. Mazaheri M, Akhavan O, Simchi A (2014) Flexible bactericidal graphene oxide–chitosan layers for stem cell proliferation. Appl Surf Sci 301:456–462
70. Liu S, Zeng TH, Hofmann M, Burcombe E, Wei J, Jiang R, Kong J, Chen Y (2011) Antibacterial activity of graphite, graphite oxide, graphene oxide, and reduced graphene oxide: membrane and oxidative stress. ACS Nano 5:6971–6980
71. Marta B, Potara M, Iliut M, Jakab E, Radu T, Imre-Lucaci F, Katona G, Popescu O, Astilean S (2015) Designing chitosan–silver nanoparticles–graphene oxide nanohybrids with enhanced antibacterial activity against *Staphylococcus aureus*. Colloids Surf, A 487:113–120
72. Jiang Y, Gong J-L, Zeng G-M, Ou X-M, Chang Y-N, Deng C-H, Zhang J, Liu H-Y, Huang S-Y (2016) Magnetic chitosan–graphene oxide composite for anti-microbial and dye removal applications. Int J Biol Macromol 82:702–710
73. Liu N, Chen X-G, Park H-J, Liu C-G, Liu C-S, Meng X-H, Yu L-J (2006) Effect of MW and concentration of chitosan on antibacterial activity of *Escherichia coli*. Carbohyd Polym 64:60–65
74. Liu H, Bao J, Du Y, Zhou X, Kennedy JF (2006) Effect of ultrasonic treatment on the biochemphysical properties of chitosan. Carbohyd Polym 64:553–559
75. Mekahlia S, Bouzid B (2009) Chitosan-copper (II) complex as antibacterial agent: synthesis, characterization and coordinating bond-activity correlation study. Phys Proc 2:1045–1053
76. Ancona A, Sportelli M, Trapani A, Picca R, Palazzo C, Bonerba E, Mezzapesa F, Tantillo G, Trapani G, Cioffi N (2014) Synthesis and characterization of hybrid copper–chitosan nano-antimicrobials by femtosecond laser-ablation in liquids. Mater Lett 136:397–400
77. Arjunan N, Singaravelu CM, Kulanthaivel J, Kandasamy J (2017) A potential photocatalytic, antimicrobial and anticancer activity of chitosan-copper nanocomposite. Int J Biol Macromol
78. Pham T-D, Lee B-K (2014) Cu doped TiO_2/GF for photocatalytic disinfection of *Escherichia coli* in bioaerosols under visible light irradiation: application and mechanism. Appl Surf Sci 296:15–23

Chapter 8
Polymers for Membrane Filtration in Water Purification

Adewale Giwa, Menatalla Ahmed and Shadi Wajih Hasan

Abstract Polymers are sometimes preferred for membrane filtration because they are more flexible, easier to handle, and less expensive than inorganic membranes fabricated from oxides, metals, and ceramics. The polymers are used as the membrane active layer and porous support in reverse osmosis (RO), nanofiltration (NF), ultrafiltration (UF), microfiltration (MF) processes. However, the application of polymers for filtration suffers critical drawbacks, such as the chemical attack of polymers, membrane fouling, and hydrophobicity of most polymers. In this chapter, the polymers used for membrane filtration in recent studies and their fabrication procedures are presented and discussed. The polymers used in recent applications include cellulose acetate (CA), polyamide (PA), polyvinylidene fluoride (PVDF), polysulfone (PSF), polyethersulfone (PES), polyvinyl chloride (PVC), polyimide (PI), polyacrylonitrile (PAN), polyethylene glycol (PEG), polyvinyl alcohol (PVA), poly(methacrylic acid) (PMAA), poly(arylene ether ketone) (PAEK), poly(ether imide) (PEI), and polyaniline nanoparticles (PANI). A new polymeric material named polyethersulfone amide (PESA) has also been presented recently. Most of the recent studies have focused on improving the specific energy consumption, salt rejection, water flux, chemical resistance and antifouling properties of polymeric membranes and nanocomposites through blending and surface modification techniques. These techniques involve the use of zwitterionic coatings, sulfonated poly (arylene ether sulfone) (SPAES), perfluorophenyl azide (PFPA), carbon nanotubes (CNTs) and graphene oxide (GO) as nanofillers, polyether ether ketone (PEEK), and nanoparticles such as titanium dioxide (TiO_2), and mesoporous silica. The use of polymers for filtration is still gaining tremendous attention, and further improvements of polymeric characteristics for enhanced membrane performance are expected in the coming years.

A. Giwa · M. Ahmed · S. W. Hasan (✉)
Department of Chemical Engineering, Khalifa University of Science and Technology, Masdar City Campus, P.O. Box 54224, Abu Dhabi, United Arab Emirates
e-mail: swajih@masdar.ac.ae

R. Das (ed.), *Polymeric Materials for Clean Water*,
Springer Series on Polymer and Composite Materials,
https://doi.org/10.1007/978-3-030-00743-0_8

8.1 Introduction

Current water quality regulations and standards require the careful treatment of water from different sources, including seawater, brackish water, groundwater, industrial wastewater, municipal wastewater, and surface water, so that the final effluents can be useful for a wide range of applications [3, 11, 12, 14]. Desalination is a key water treatment approach, most especially in areas with inadequate natural and renewable fresh water supply. In addition, the industries in many countries are now being asked to treat their own wastewater to reduce dependence on naturally available fresh water and desalinated water [21, 59]. However, treated water from industries and municipalities require an appreciable level of treatment to prevent possible secondary or end-use problems.

Conventionally, treatment processes such as biological treatment, distillation, evaporation, chemical coagulation, flocculation, sand filtration, and gravity sedimentation are used to remove pollutants from water [31, 35, 38]. For saline water desalination, thermal distillation processes such as multi-effect distillation, multi-stage flash, and thermal and mechanical vapor compression were mainly employed in parts of the world until the recent decades. Nowadays, desalination via membrane filtration processes, i.e., microfiltration (MF), ultrafiltration (UF), nanofiltration (NF), and reverse osmosis, (RO) is gaining immense attention from desalination stakeholders [6, 88, 100]. Currently, the market share of membrane filtration processes in the desalination industry has soared and surpassed those of other approaches. MF and UF are mostly being used for pre-treatment in current desalination applications, instead of coagulation or sedimentation. In addition, there are more RO plants in the world than those that employ thermal desalination approaches currently [28]. Likewise, membrane filtration processes such as membrane bioreactors (MBRs) and osmotic membrane bioreactors (OMBRs) are being preferred for wastewater treatment than conventional approaches such as activated sludge processes (ASP), aerated lagoons, and trickling filters [17, 33].

The functional component of a membrane filtration process is the membrane. Membranes can be made from polymeric or inorganic materials [66, 104]. Most of the polymeric membranes are organic in nature, while inorganic membranes are mainly oxides, ceramics, and metals [89]. Membranes made from polymeric materials are cheaper than those fabricated from inorganic materials or ceramics [66]. Additionally, polymeric membranes can be used to achieve high water production capacity. These membranes are easy to handle during fabrication and can be arranged in different configurations such as hollow fiber and spiral wound for optimum performance [46, 66, 89]. Therefore, the objective of this chapter is to review the polymeric materials that have been used for filtration in recent times. Specifically, the recent advances in the fabrication of polymers for RO, NF, UF, and MF processes are discussed. The type of polymer used for filtration is crucial because it determines the permeate quality and the operating cost of water production. Proper selection of polymer is required for a filtration process to ensure that issues such as frequent membrane replacement and unwarranted energy consumption are avoided. The current challenges associated with recently devised polymers are also presented and discussed.

8.2 Polymers Used for Membrane Filtration

Several polymers have been used in the fabrication of MF, UF, NF, and RO membranes. Examples include cellulose acetate (CA), polyamide (PA), polyvinylidene fluoride (PVDF), polysulfone (PSF), polyethersulfone (PES), polyvinyl chloride (PVC), polyimide (PI), polyacrylonitrile (PAN), polyethylene glycol (PEG), polyvinyl alcohol (PVA), poly(methacrylic acid) (PMAA), poly (arylene ether ketone) (PAEK), poly(arylene ether sulfone) (PAES), poly(ether imide) (PEI), and polyaniline nanoparticles (PANI). A new polymeric material named polyethersulfone amide (PESA) has also been presented recently. Fabrications, characterization, and related applications of such membranes are highlighted in the subsequent sections.

8.2.1 Polymers for RO

RO technology (Fig. 8.1) has been found to be one of the most efficient and widely popular methods of desalinating water because it is suitable for the production of potable and near-to-potable water [41, 53].

RO membranes that are commercially available consist of polymeric materials such as CA and PA [13, 28]. CA is used because it is a natural polymer that is renewable, biodegradable, and eco-friendly [18, 63]. CA can be produced through the esterification of wood, cotton, recycled paper, and bagasse. CA is also a widely used polymer known for its high hydrophilicity, biocompatibility, high potential flux, etc. [26]. However, PA membranes are generally preferred among the two because of their ability to operate under a wider pH range and withstand higher temperatures [72]. Unfortunately, the practical application of PA membranes is often limited due to their continuous exposure to chlorine and other oxidizing substances [110]. The amide group that is present in the PA membranes is vulnerable to chlorine attacks during chemical cleaning [28, 92]. Hence, an additional de-chlorination step is required to reduce the concentration of chlorine to prevent the degradation of the PA membranes. Also, in order to overcome this problem, poly(arylene ether) copolymers, especially poly(arylene ether sulfone), have been

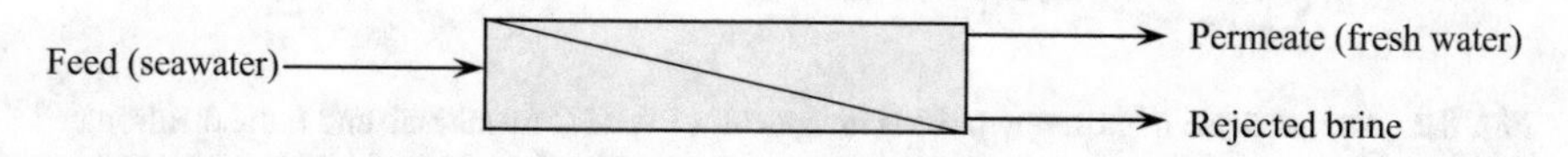

Fig. 8.1 RO process showing the separation of salt from water

used recently for RO desalination [67, 68, 92]. Since these polymers do not contain any susceptible amide linkages, it makes them to be highly resistant to chlorine attacks. It has also been established recently that thin film composite (TFC) membranes based on SPAES display high chlorine-tolerance and no significant change in water flux after 36 h of continuous exposure [110].

(a) poly(arylene ether sulfone) (PAES)

Photocross-linkable disulfonated PAES copolymers have been prepared for RO applications in a recent study [64]. First, PAES oligomers with controlled degrees of sulfonation and controlled molecular weights were synthesized via nucleophilic aromatic substitution. Meta-aminophenol was used to control the molecular weight of the PAES oligomers and install telechelic amine end groups. The meta-aminophenol end-capped oligomers were reacted with acryloyl chloride to obtain novel cross-linkable PAES oligomers with acrylamide groups on both ends. The acrylamide-terminated oligomers were cross-linked using UV radiation in the presence of a multifunctional acrylate and a UV photoinitiator to obtain PAES copolymers thin films. It was shown that the cross-linked disulfonated PAES films had smooth surfaces that promoted high water passage (Fig. 8.2). The copolymer films also exhibited reduced water uptake and swelling relative to their linear counterparts.

(b) Polyamide (PA)

Apart from the problem of chlorine attacks during chemical cleaning, PA polymeric material faces membrane fouling [19, 78, 79]. Biofouling is one of the most challenging fouling mechanisms experienced during membrane filtration [5, 39, 44]. Biofouling occurs due to the formation of a biofilm by the biological species in a membrane filtration system, resulting in the depletion of the membrane's lifetime and selectivity. Although RO works based on the solution–diffusion principle rather than size exclusion, biofouling is a major problem in RO. This can be attributed to the thin layer of the active surface and the material of the dense and porous structures [54, 90]. Thus, zwitterionic natured coatings on membranes have been observed to be effective antibiofouling materials in recent studies

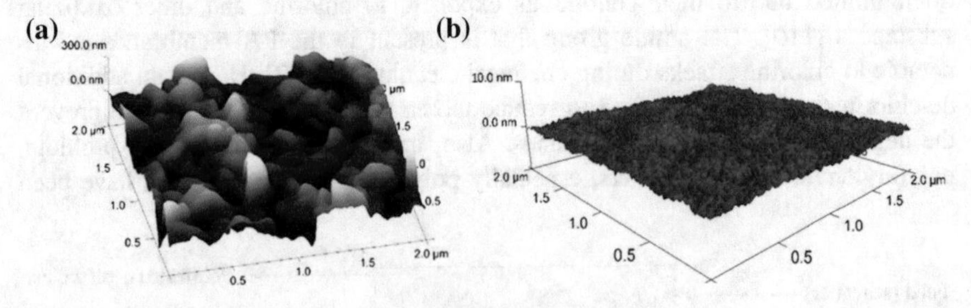

Fig. 8.2 Atomic force microscopy (AFM) images of **a** PA TFC membrane and **b** the disulfonated copolymer, showing the smooth surface of the copolymer film that ensured higher water passage [64]

[50, 80]. A desirable feature of the zwitterionic structure is that it exhibits both positively and negatively charged moieties within the same segment side chain which imparts strong hydration capacity via electrostatic interactions. These polyzwitterions are usually attached to the membrane's surface using a grafting approach. The coatings produce densely packed polymer chains that exhibit consistent length and reduce the adhesion of cells and bacteria onto the membrane surface [27]. The surface modification of polymeric RO membrane by zwitterionic polymer can be used to achieve higher permselectivity and water flux [96]. The modification has been accomplished recently by the grafting of a commercially available membrane with N,N′dimethylaminoethyl methacrylate (DMAEMA) via redox-initiated graft polymerization reaction [96]. Then, the DMAEMA graft was modified via surface quaternization reaction with 3-bromopropionic acid (3-BPA) to obtain the zwitterionic carboxybetaine methacrylate (CBMA) polymer chains on the membrane surface. The CBMA, which has a cationic quaternary ammonium group and anionic carboxylate group on its backbone, changed the chemical structure, morphological structure, hydrophilicity, and charge of the RO membrane. The fabricated procedures are illustrated in Fig. 8.3. The modified membrane showed improved water flux (22.5% increase in water flux compared to the unmodified membrane). By using positively charged lysozyme and negatively charged bovine serum albumin as foulants, it was shown that the biofouling properties of the modified membrane were enhanced as evidenced by the higher water recovery rate after fouling test. Up to 99% mortality of *Escherichia coli* (*E.coli*) and *Bacillus subtilis* was achieved. Another polymer that has been used successfully for grafting is polysulfobetaine. Polysulfobetaine was grafted from the surface of commercially available TFC membranes in another recent study [27], leading to 80% reduction in microbial fouling without any adverse effect on the permeate flux.

PA-RO membranes with enhanced antifouling properties have also been prepared recently by making use of the highly reactive azide group of PFPA that can form chemical bonds when activated by photoexcitation with nonreactive groups [57]. First, PEG polymers were modified with a terminal PFPA group. Then, pieces of commercially available PA were dipped into an aqueous solution containing the PEG-PFPA prepared polymers. The pieces of modified PA were dried at ambient conditions and irradiated with 254 nm UV light. Finally, the pieces of PA obtained were rinsed with water to remove unreacted azides and other by-products. The performance of the prepared membranes was evaluated through pure water permeability and sodium chloride (NaCl) rejection tests. The antibiofouling properties of the membranes were assessed by monitoring the growth of *E.coli* on the membranes. The prepared membranes were more hydrophilic than the commercially available PA. The membranes also exhibited lower water permeability but increased NaCl rejection. It was observed that the prepared membranes had better antibiofouling properties as evidenced by the reduced growth of *E.coli* bacteria on the prepared membranes.

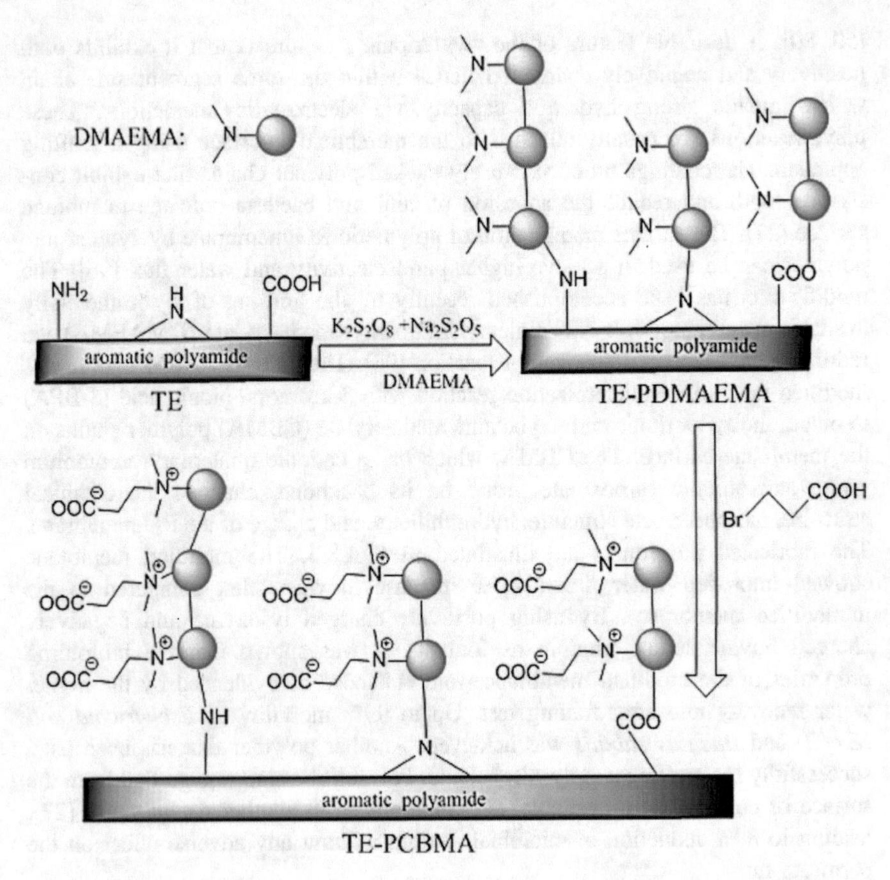

Fig. 8.3 Illustration of the fabrication procedures for the modification of RO membranes with zwitterionic polymer chains [96]

(c) Polyvinyl chloride (PVC)

Another polymeric material that has been used recently for the RO process is polyvinyl chloride (PVC) [2, 70]. This is because of its flexibility and durability along with suitable biological and chemical resistance [26]. Special selective characteristics and enhancement of separation properties have been achieved in membranes with the use of PVC/CA polymers as membrane binders. It has been observed that an increase in CA concentration in the dope solution that consists of PVC/CA polymers would result in an increase in the hydrophilic characteristics of the membrane [26]. This is because the high amount of water would be absorbed; hence, more water would pass through the membranes. An increase of CA concentration to about 10% could also improve the rejection capabilities of the fabricated membrane.

(d) Chemical modifications of membrane properties using carbon nanotubes (CNTs)

Other studies have also shown that membrane properties can be improved when polymeric materials are chemically modified with other polymers of desirable properties [80, 86]. One of such additives is carbon nanotubes (CNTs) which act as nanofiller in RO desalination. Although a decrease in the membrane permeability would be observed when the CNT concentration is increased, an increase in the salt rejection and permeate flux would be achieved [77]. CA is efficient in the rejection of salts during RO desalination because of its excellent desalting properties resulting from its nanoscale characteristics [16]. Recently, raw and oxidized multiwalled CNTs (MWCNTs) in different concentrations (0.001, 0.002, 0.005, 0.01 wt %) have been incorporated into PA-RO membranes [23]. The morphology of the modified membranes was altered as a result of the MWCNTs incorporation. The membranes embedded with raw MWCNTs exhibited slightly higher contact angle compared to the pristine membrane, while membranes embedded with oxidized MWCNTs had slightly lower contact angle compared to the pristine membrane. An increase in the concentration of both raw and oxidized MWCNTs up to 0.005 wt% resulted in an increase in the water flux, after which the water flux decreased. Meanwhile, all concentrations of raw or oxidized MWCNTs resulted in better antifouling performance of the modified membranes. The modified membranes with 0.005 wt% MWCNT concentration showed the best antifouling properties.

Kim et al. [42] have also demonstrated recently that the modification of PA-RO membrane with CNT can be used to accomplish improved membrane properties [42]. The CNTs were initially functionalized by reacting them with a sulfuric acid/nitric acid mixture. Then, PA was prepared by using trimesoyl chloride (TMC) solutions in n-hexane and aqueous solutions of m-phenylenediamine (MPD) containing the functionalized CNTs. The maximum flux and salt rejection values were observed when the functionalized CNTs were prepared by the reactions of CNTs with a sulfuric acid and nitric acid mixture for 4 h at 65 °C. When shorter reaction time and lower reaction temperature were used, the CNTs were not well-dispersed in the PA active layers. Conversely, when longer reaction time and higher reaction temperature were used, the CNTs were cut down into very small pieces to form aggregated structures. Therefore, good dispersion of the functionalized CNTs in the PA layer was necessary. The membranes containing the properly modified CNTs demonstrated higher water flux than the PA membrane prepared without any CNTs. Better chemical resistance against NaCl solution compared to the pristine RO membrane was also achieved by using the modified membranes.

However, the mechanical strength and structural integrity of the nanofiller still need to be improved in future research activities so that it can be employed for large-scale commercial desalination. To achieve mechanical stability, a recent study has tested 1,2-bis(triethoxysilyl)ethane (BTESE) instead. In this study [34], a porous PSF-supported BTESE hybrid membrane was fabricated through a sol–gel spin-coating heat treatment process. A 200-nm-thick BTESE-derived silica separation layer was deposited onto the PSF support surface. The RO membrane was evaluated by using it to desalinate a NaCl aqueous solution. The membrane showed a stable and high degree of water permeability with high salt rejection reaching 96%. The membrane also showed good stability and reproducibility during the RO desalination process that was run for more than 160 h.

8.2.2 Polymers for NF

(a) Polyimide (PI) and polyamide (PA)

NF membranes have gained popularity for water filtration in recent decades due to their beneficial features such as low energy consumption when compared with RO and high retention of divalent salts and neutral molecules of low molecular weight [25, 61, 112]. Nonetheless, NF membranes can only withstand aqueous solutions containing pH in the range of 2–11 due to their moderate stability. Most of the NF membranes available today consist of PI, PA, PVA, and PAN polymers in TFCs [4, 85, 87, 93, 95, 105, 108]. However, PIs are unstable when in contact with a few amines. They also exhibit very low stability and performance in polar solvents. These PIs are not preferred in aqueous solutions containing chlorinated solvents, strong amines, and strong acids/bases, but they can be modified through the process of cross-linking to obtain improved resistance against such chemicals.

(b) Poly(arylene ether ketone) (PAEK)

An alternative solution that involves the use of PEEK as NF membrane material has been proposed recently [15]. It was observed that PEEK membranes have a low degree of sulfonation and are highly resistant against various solvents, acids, and bases. However, PEEK membranes exhibit low water permeability. The PEEK membranes were tested for their separation performance in tetrahydrofuran (THF) and dimethylformamide (DMF) where they exhibited a water permeance of 0.2–0.8 and 0.7–0.21 L/h m^2 bar, respectively.

(c) Membrane fouling in NF membranes

The challenge of membrane fouling and chemical attack is also associated with PA NF membranes. Fouling not only reduces the flux through NF membranes but also increases the energy requirement. Meanwhile, surface modification has been employed recently to impart antifouling properties to PA NF membranes [48, 52, 60, 109]. These properties have been achieved by grafting fluorinated PA onto the surface of the PA NF membranes [48]. The fluorinated PA NF membranes have lower surface energy which resulted in the minimization of the adhesion propensity membrane. The detachment of foulants from the membrane surface was achieved through the fluorination of the membrane. 98.3% permeation flux recovery was accomplished through this approach. Stability problems are also associated with PVA NF membranes. However, in a recent investigation, a novel TFC membrane has been fabricated by cross-linking PVA and 3-mercaptopropyltriethoxysilane on porous PSF support in order to enhance the ion rejection and acid/alkali stability of the membrane [109]. The introduction of a sulfonic acid group enhances the hydrophilic properties of the membrane which in turn caused an increase in the water flux across the membrane. This approach is not completely advantageous because the sulfonic acid groups also caused the swelling of the membrane, resulting in a decrease in the membrane's rejection properties.

Meanwhile, antifouling and salt rejection features can be imparted to a TFC NF membrane by replacing the mid-layer of the TFC membrane with an electrospun nanofibrous membrane (ENM) resulting in a TFNC membrane [84]. ENMs are known for their large dirt loading capacity due to their large internal surface area. In order to achieve this, the ENM layer must be hydrophilic and heat-treated before interfacial polymerization. In addition, they are highly porous compared to conventional membranes which would ensure that the water flux across the membrane is enhanced. ENMs are produced through electrospinning and have unique properties such as high surface area to volume ratio, tailorable pore sizes, and flexibility in their surface chemistry.

8.2.3 Polymers for UF

PS and PES are widely used in UF membranes because they are polymeric materials with good mechanical properties, wide pH operation range, and strong chemical stability [20, 62, 69, 82, 83, 91, 101]. However, their application in water treatment is limited due to their hydrophobicity which ultimately leads to reduced membrane permeability. Most of the polymeric membrane materials that are widely used in UF processes exhibit hydrophobic properties. PVDF, PVC, and PMAA have also been used recently for the fabrication of UF membranes. These polymers are also naturally hydrophobic [10, 37, 55, 94 102, 111]. Membrane hydrophobicity can cause water flux decline during operation due to the accumulation of organic compounds that favor the attachment and growth of microorganisms onto the membrane surface. This usually leads to membrane fouling and subsequently membrane failure. Thus, to improve their properties and enhance their performance in water treatment applications, modifications to these polymeric materials are necessary. These modifications are carried out in such a way that the membrane hydrophilicity is increased. An increase in the membrane surface hydrophilicity would enhance the membrane's antifouling properties for liquid water-based filtration.

(a) Incorporation of TiO_2 nanoparticles into polysulfone (PSF)

Blending and surface modification can be used to incorporate hydrophilic materials (nanoparticles and amphiphilic copolymers) into UF membranes to increase their hydrophilicity [49]. A hybrid PSF membrane impregnated with modified TiO_2 nanoparticles for the impartation of hydrophilic property to PSF has been proposed recently [66, 107]. The membrane was prepared by grafting the hydrophilic polymer chains of (2-hydroxyethyl methacrylate) (P(HEMA)) on TiO_2 nanoparticles through atom transfer radical polymerization process. PSF membranes were impregnated with the modified TiO_2 nanoparticles for achieving better membrane performance, overcoming agglomeration of nanoparticles on the membrane surface and reducing the leakage of nanoparticles from the membrane during filtration. The modified TiO_2 particles had better dispersibility within the polymer than unmodified TiO_2. The PSF membrane modified with TiO_2-HEMA exhibited improved hydrophilicity, higher water flux, and better antifouling performance than the pristine PSF membrane and unmodified TiO_2 impregnated membranes.

(b) Incorporation of mesoporous silica particles (MSP-1) into polysulfone (PSF)

Incorporating inorganic particles into a membrane's casting mixture prior to phase inversion is widely studied because it is a facile approach that can be used to embed additional particle-based functionalities into membranes [16, 29, 73]. Surfactant-templated mesoporous silica particles (MSP-1) have been incorporated into PSF matrices formed with and without PEG as a molecular porogen with the aim of enhancing the properties of PSF membranes [22]. It was observed that MSP-1 additives increased the hydrophilicity of the membrane by virtue of the terminal silanol (Si–OH) groups on the pore walls and external surfaces of the particles. Both MSP-1 and PEG modified the typical morphology of the phase inversion membrane content. The mechanical properties of the PSF–MSP mesocomposite were comparable to those of their MSP-free counterparts. The addition of MSP-1 to porogen-free membranes made from casting solutions with low polymer content led to statistically significant differences in permeate flux. The addition of only 5.0 wt% MSP-1 had a detrimental effect on flux, yet a further increase to 10 wt% loading level raised the permeate flux above the value observed for MSP-free controls. However, when the PEG porogen was included in the casting mixture, no statistically significant changes either in flux or in rejection were observed. The mesocomposite membranes showed enhanced dextran rejection compared to MSP-free membranes, and fouling tests with humic acid solutions demonstrated that the mesocomposite membranes experienced lower flux decline and showed higher rejections than their MSP-free counterparts.

(c) Incorporation of zinc oxide (ZnO) and silica nanoparticles into polyvinyl chloride (PVC) and poly(methacrylic acid) (PMAA)

The impact of incorporating nanoparticles into PVC for enhanced hydrophilicity of PVC UF membranes has also been studied recently [75]. The effect of incorporating zinc oxide (ZnO) nanoparticles into PVC membranes was examined. Five PVC membranes having variable ZnO percentages (0.3, 1.0, 2.0, 3.0, and 4.0 wt%) were fabricated via the phase inversion method using water as coagulant and PEG as a pore forming additive. The ZnO impregnated membranes had a higher hydrophilicity than pristine PVC membranes, with the 4.0 wt%-ZnO membrane being the most hydrophilic. An increase in ZnO concentration up to 3.0 wt% led to an increase in water flux. Further increase in ZnO concentration led to a decline in water flux due to the agglomeration of ZnO particles at the surface of the PVC membrane (Fig. 8.4). An increase in ZnO concentration up to 3.0 wt% also led to increase in membrane porosity, after which it declined. The pristine PVC membranes were only able to recover 69% of water flux after BSA permeation, whereas membranes containing 3.0 wt% ZnO were able to recover 92% of water flux after BSA permeation. Incorporation of nanoparticles into PMAA has also been examined recently for the improvement of the performance of PMAA UF membranes. Superhydrophilic silica nanoparticles have been grafted onto PMAA membranes through the process of post-fabrication tethering [49]. An increase in the wettability of the membranes was observed.

(d) Polyethersulfone (PES)

PES has been used in most of the recent studies on UF membrane separation. The hydrophilicity of PES membrane has been improved recently by incorporating mesostructured silica particles functionalized with amine and carboxylic groups into PES [56]. The morphology, porosity, and pore size distribution of the modified membrane changed significantly as a result of the incorporation of ordered mesoporous silica particles. The hydrophilicity of the modified membrane also increased significantly. Water permeation through the membrane increased as a result of the enhanced surface porosity and hydrophilicity of the modified membrane. The antifouling property of the modified membrane was improved, especially against irreversible fouling, without negatively affecting the protein rejection potential of the membrane. It was also observed that the modified membrane exhibited a stable permeation performance during repeated stability tests. In another recent study, a new hydrophilic polymeric material that is based on PES has been proposed. This material was named polyether sulfone amide (PESA) [58]. PESA was prepared through the polycondensation reaction of diamine (4,40-diaminodiphenyl ether) with dicarboxylic acid (diphenyl sulfone 4,40-dicarboxylic acid) using triphenyl phosphite (TTP), lithium chloride (LiCl), calcium chloride ($CaCl_2$), and pyridine (Py) as condensing agents and N-methyl-2-pyrrolidone (NMP) as a solvent. PESA was further modified by grafting it with two hydrophilic monomers, i.e., 3,5-diaminobenzoic acid (DBA) and gallic acid (GA) via interfacial polymerization. It was observed that PESA membrane was more hydrophilic than pure PES membrane. The modification of PESA membrane with DBA and GA further increased the membrane's hydrophilicity. PESA membrane and modified PESA membrane had greater roughness compared to pure PES membranes. The pure water flux and humic acid rejection of PESA membrane were higher than those of the pristine PES membrane. PESA membrane also showed higher antifouling properties than the PES membrane. The antifouling properties of PESA membranes were further improved by surface modification with DBA and GA.

The hydrophilicity of PES membrane has also been enhanced by incorporating PANI nanoparticles into PES UF membranes [48, 76]. To do this, three different

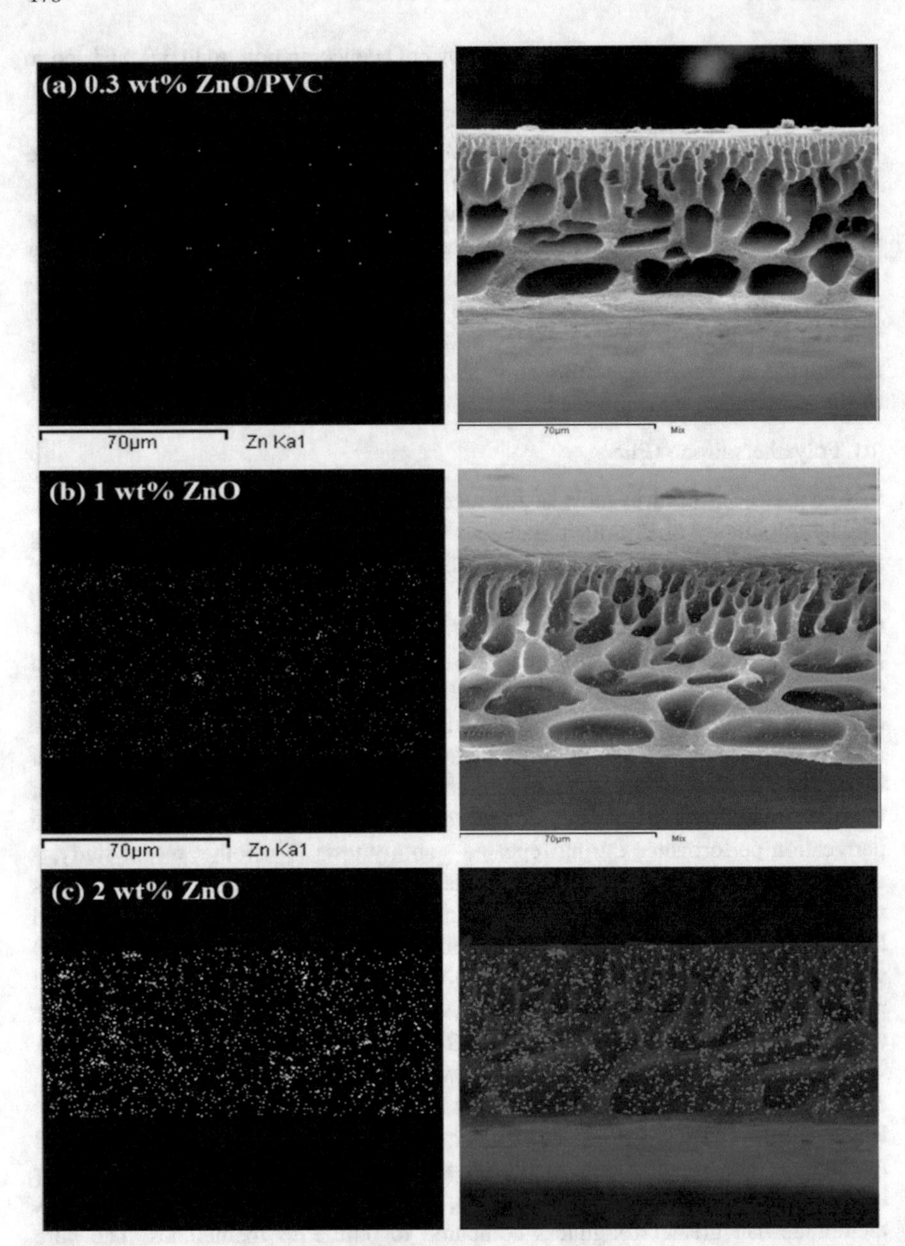

Fig. 8.4 Energy dispersive X-ray (EDAX) spectroscopy showing agglomeration as more ZnO particles are included in the PVC-ZnO casting solution [75]

membranes—pure polyethersulfone, self-synthesized PANI impregnated into PES, and commercially available PANI impregnated into PES—were fabricated via phase inversion. The membranes were characterized via contact angle goniometry and evaluated through direct interaction with BSA, humic acid, silica nanoparticles,

E.coli and *Bacillus* bacteria. The addition of PANI nanoparticles led to increased hydrophilicity, enhanced fouling resistance, better flux recovery, improved BSA and humic acid rejection, and reduced attack from bacteria. Interestingly, the self-synthesized PANI impregnated into PES membrane was superior to the commercially available PANI impregnated into PES membrane, in terms of membrane properties. PES UF membranes have also been modified by incorporating highly hydrophilic polyethylene glycol (PEG) and silver nanoparticles (Ag) into PES using poly(acrylonitrile-co-maleic acid) (PANCMA) as a chemical linker [71]. Hollow fiber configuration was used. Polymeric membranes with hollow fiber configuration are preferred for some separation processes because this configuration has the advantages of high surface area, self-mechanical support, excellent flexibility, and ease of handling during module fabrication. The modified membrane was shown to exhibit enhanced properties including higher hydrophilicity (75.5% decrease in contact angle), increased water flux (by 36%), and reduced bacterial growth. Another recent study has improved the hydrophilicity and antifouling property of PES UF membrane by modifying it with dextran-grafted halloysite nanotubes (HNTs) [106]. The incorporation of dextran-HNTs into PES membranes led to significant increase in hydrophilicity as evidenced by the reduction of water contact angle. In addition, the modified membranes showed higher flux and better antifouling properties than pristine PES membranes. Interestingly, the modified membranes had a slightly lower porosity, yet larger pore size than the pure PES membranes.

Meanwhile, the conventional multi-bore hollow fiber membrane consists of three or seven bore channels and an outer round-shaped geometry. However, the main drawback of this geometry is the nonuniform wall thickness. The thinner part of the membrane wall suffers as the mechanically weak point, while the thicker part generates additional mass transfer resistance. Therefore, to overcome this drawback, an attempt has been recently directed toward the fabrication of a novel tri-bore hollow fiber membrane with round-shaped bore channels but an outer triangle-shaped geometry made of Matrimids and PES materials [97]. The triangle-shaped tri-bore hollow fibers can be fabricated with a combination of a tri-bore blossom spinneret and defined spinning parameters. The new geometry, which exhibits a much more uniform wall thickness, was shown to improve the mechanical properties of both the Matrimids and PES membranes as well as their water permeation.

(e) Polyvinylidene fluoride (PVDF)

PVDF is another polymeric material that can be modified for enhanced hydrophilicity. PVDF UF membrane has been recently modified by dipping the PVDF membrane into a dopamine solution such that the dopamine coats the surface of the PVDF membrane by self-polymerization [81]. The coated PVDF membrane was rinsed with water to remove unreacted polydopamine. The coated PVDF membrane obtained was then dipped into a solution containing TiO_2 nanoparticles. The dopamine acted as a glue to facilitate the attachment and distribution of the

TiO_2 nanoparticles onto the PVDF membrane. The resulting PVDF membrane was rinsed with water to remove large TiO_2 particles that are deposited onto the surface of the polydopamine-coated PVDF membrane. The TiO_2 nanoparticles were homogeneously distributed on the surface of PVDF and did not agglomerate. The hydrophilicity of the modified membrane was improved as evident in the significant reduction of the water contact angle of the membrane. The pure water flux across the modified membrane also increased significantly, and the BSA rejection of the membrane was enhanced. The antifouling properties of the membrane were improved as evident in the low irreversible fouling ratio and a remarkably high flux recovery ratio (>90%) achieved for BSA separation.

(f) Poly(arylene ether ketone) (PAEK) and poly(ether imide) (PEI)

Other polymeric membranes that have been tested for UF operations in recent studies are PAEK and PEI [40, 51]. Cardo PAEK membrane bearing hydrophilic carboxylic acid groups (PAEK-COOH) has been proposed as an alternative to the traditional hydrophobic PAEK membranes [51]. PAEK with pendent carboxylic acid group (PAEK-COOH) was first synthesized by the aromatic nucleophilic substitution polycondensation reaction of 2-[bis(4-hydroxyphenyl)methyl] benzoic acid (PPH-COOH) and 4,4′-bisfluorodiphenylketone in DMSO. Thermal analyses demonstrated that the synthesized PAEK-COOH polymer has a decomposition temperature of 360 °C and glass transition temperature of 220 °C, which suggests that it is well qualified for preparing membranes to deal with hot water without temperature controlling. Then, the synthesized polymer was used to prepare a tight UF membrane by the nonsolvent-induced phase inversion process. The resulting membrane had a water contact angle of 61.5°. The membrane displayed high water permeation flux and dye rejection. The antifouling performance and antidye adsorption properties of the membrane are also promising, possessing a flux recovery ratio of 91.5% for BSA, and dye adsorption rate below 5.0% for all the studied dyes (Congo red, Coomassie brilliant blue R250, Direct red 23, and Evans blue (EB). The membrane is thermally stable and suitable for high-temperature filtration applications.

PEI UF membrane has been modified recently by blending PEI with N-phthaloylated chitosan (NPHC) so as to enhance the antifouling properties of the membrane [40]. The modified membrane was more hydrophilic than the unmodified membrane. The roughness of the surface of the modified membrane was greater than that of the unmodified membrane. The surface roughness increased with increasing NPHCs content. Pure water flux increased with increasing concentration of NPHCs in the NPHCs blended membrane. Meanwhile, when the concentration of NPHCs in the NPHCs blended membrane was increased, the capacity of the fabricated membrane to reject protein became lower while permeate flux increased. However, the separation of heavy metal ions increased with increasing concentration of NPHCs in the NPHCs blended membrane. Maximum flux recovery was achieved for the PEI/NPHCs blended membrane when the NPHCs concentration was 2.0 wt%.

8.2.4 Polymers for MF

MF membranes have been mainly used in membrane distillation (MD), MBRs, and wastewater treatment processes [1, 7, 30, 32, 99]. Industrial and domestic wastewater contains harmful organic pollutants (like pharmaceutical compounds) which constitute a great threat to aquatic species and the environment in general. Advanced technologies such as advanced oxidation processes (AOPs) have proven to be very efficient in cleaning recalcitrant wastewater. One of the most important AOPs is photocatalysis via TiO_2, which completely mineralizes a wide range of organic compounds. The direct incorporation of TiO_2 nanoparticles onto MF polymer membranes has been proven to be a viable membrane separation technique recently [24]. Titanium tetraisopropoxide (TTIP) was used as the precursor for TiO_2. TTIP hydrolysis prevented the formation of agglomerates and increased the bonding strength of the TiO_2 particles formed. PES and PVDF membranes were used as the supporting structures for TiO_2 nanoparticles. After the attachment of TiO_2 particles, a decrease in the porosity of the membrane was observed. However, the attached TiO_2 showed the ability to degrade various molecules like dyes, drugs, and pesticides.

MD is a thermally driven process by which water molecules are separated from other undesired substances through porous membranes [7, 8]. Hydrophobic membranes are required for MD applications because MD works based on the principle of vapor permeation. The vapor pressure difference across a hydrophobic membrane is the driving force in MD. Although MD is known for its easy implementation and utilization of heat, it has not yet gained industrial-scale application due to its drawbacks, among which MF membrane fouling and low flux are the most predominant and hard to tackle. Nonetheless, it has recently been reported that an increase in the flux had been observed in DCMD membranes by tetrafluoromethane (CF4) plasma surface modification [103]. Although the vapor flux through the plasma-modified membrane reached its maximum at about 15 min and then started to decrease, the overall flux of the modified membrane was still higher than that of the virgin membrane. The PVDF membranes were converted to superhydrophobic membranes through CF4 plasma treatment, which resulted in the enhancement of flux and salt rejection.

MBRs have gained popularity in wastewater treatment due to its high quality of processed water, reduction in excess sludge, controllability of solids, and minimization of required footprint [36, 65]. Although it has many positives, one of the most important drawbacks of MBR operations is also membrane fouling [9, 47]. In order to potentially overcome this, graphene oxide (GO) is currently being incorporated with MF membranes to prepare MBR membranes with antifouling properties [45]. This is due to the unique properties of GO such as hydrophilicity and large negative zeta-potential attributable to its functional groups. These properties enhance water permeation through the membrane and impede biofouling. It has been observed that the thickness of biofilm formed by the microorganism on GO-incorporated membranes decreases and the negative zeta-potential increases when the GO content within the membrane is increased [47]. The addition (up to 1.3 wt%) of GO to the membrane dope helped to prevent fouling and increase pure water flux through the membrane significantly. Above 1.3 wt% of GO would result in an increase in polymer solution viscosity, which can result in the reduction of the membrane pore size and water flux. High energy demands resulting from membrane fouling is an indirect drawback associated with MBR operations. Therefore, research on osmotic MBRs (OMBRs) has been intensified recently [43, 53, 98]. The driving force in OMBRs is the osmotic pressure difference between the feed and draw sides of the membrane, rather than hydraulic pressure difference. However, as compared to a conventional MBRs, OMBRs contain a high rejection semipermeable membrane instead of a microporous membrane. Although the fouling potential is comparatively much lower in OMBRs, membrane fouling still does occur. The elevated salinity and salt accumulation, the interactions of inorganic ions and organic foulants, and the scaling of low soluble salts under high ionic strengths might even contribute toward more complex fouling phenomena [74]. But, due to the absence of hydraulic pressure in OMBRs, the compaction of membrane foulants is milder and hence fouling could be easily curtailed by hydrodynamic shear.

8.3 Summary

The fabrication procedures, features, and performance of the polymers used in recent filtration processes are discussed in this chapter. These processes include RO, NF, UF, and MF. The polymers used recently in systems such as MBRs and MD, where membrane filtration is employed, are also discussed. The efficiency of a membrane filtration process depends on the type of polymer, the physical characteristics of a polymeric membrane, and the functional groups on the surface and interior of the polymeric membrane. A membrane with desirable properties can be fabricated through the modification of the membrane-forming polymers. This modification can be achieved through the incorporation of materials such as copolymers and nanoparticles into the membrane matrix. For RO, there is a general preference for PA TFC membranes recently, but the application of these membranes is restricted by the chemical attack of the amide group in PA due to chlorine and other oxidizing compounds. Meanwhile, surface modification by zwitterionic polymer can be used to achieve higher permselectivity and water flux. In addition, SPAES lacks amide linkages; so TFC membranes based on SPAES have high resistance to chlorine attacks. These zwitterionic coatings and reactive groups that can form chemical bonds such as PFPA also have the potential to impart antifouling properties to RO membranes. Nanofillers such as CNTs can be employed to improve the salt rejection properties of RO membranes. However, further research is needed to improve the mechanical integrity of CNT nanofillers for long-term processes. For NF, PEEK contains a low level of sulfonation and can be used with polymeric membranes to achieve resistance again solvents, acids, and bases. The grafting of fluorinated PA onto the surface of the PA NF membranes could be used to improve the resistance of PA TFC membranes to fouling. The fluorinated PA has the potential to reduce the surface energy of TFC membranes, thereby reducing the adsorption of foulants on the membranes. In addition, cross-linkers and sulfonic acid group are capable of enhancing the hydrophilicity of NF membranes.

Most of the recent works on polymers used for UF have been focused on the incorporation of inorganic particles into the polymeric casting mixture prior to phase inversion. The use of surfactant-templated mesoporous silica particles, GO, and ZnO in the fabrication of UF membranes is capable of improving the hydrophilicity of the nanocomposites formed. An improvement in membrane hydrophilicity might result in an increase in pure water flux across the membrane and membrane fouling reduction. The separation properties of the nanocomposites such as morphology, porosity, and pore size distribution can be significantly tailored through such modifications. The dipping of the PVDF membrane into a dopamine solution has also been shown as a method of imparting hydrophilicity to PVDF UF membrane in a recent investigation. Recent advancements in the modifications of the functional and structural properties of polymers for filtration are ongoing, and it is expected that further improvements in the future would ensure more efficient and less expensive filtration processes.

References and Future Readings

1. Abdel-Karim A, Gad-Allah TA, El-Kalliny AS, Ahmed SIA, Souaya ER, Badawy MI, Ulbricht M (2017) Fabrication of modified polyethersulfone membranes for wastewater treatment by submerged membrane bioreactor. Sep Purif Technol 175:36–46. https://doi.org/10.1016/j.seppur.2016.10.060
2. Ahmad A, Jamshaid F, Adrees M, Iqbal SS, Sabir A, Riaz T, Zaheer H, Islam A, Jamil T (2017) Novel polyurethane/polyvinyl chloride-co-vinyl acetate crosslinked membrane for reverse osmosis (RO). Desalination 420:136–144. https://doi.org/10.1016/j.desal.2017.07.007
3. Ahmad R (2017) Water worldwide—US water regulations and India's water challenges. J Am Water Works Assoc 109:64–67. https://doi.org/10.5942/jawwa.2017.109.0041
4. Ahmed FE, Lalia BS, Hilal N, Hashaikeh R (2017) Electrically conducting nanofiltration membranes based on networked cellulose and carbon nanostructures. Desalination 406:60–66. https://doi.org/10.1016/j.desal.2016.09.005
5. Al Ashhab A, Gillor O, Herzberg M (2014) Biofouling of reverse-osmosis membranes under different shear rates during tertiary wastewater desalination: microbial community composition. Water Res 67:86–95. https://doi.org/10.1016/j.watres.2014.09.007
6. Amy G, Ghaffour N, Li Z, Francis L, Linares RV, Missimer T, Lattemann S (2017) Membrane-based seawater desalination: present and future prospects. Desalination 401:16–21. https://doi.org/10.1016/j.desal.2016.10.002
7. Ashoor BB, Mansour S, Giwa A, Dufour V, Hasan SW (2016) Principles and applications of direct contact membrane distillation (DCMD): a comprehensive review. Desalination 398:222–246. https://doi.org/10.1016/j.desal.2016.07.043
8. Ashoor BB, Fath H, Marquardt W, Mhamdi A (2012) Dynamic modeling of direct contact membrane distillation processes. Comput Aided Chem Eng 170–174. https://doi.org/10.1016/b978-0-444-59507-2.50026-3

9. Aslam M, Charfi A, Lesage G, Heran M, Kim J (2017) Membrane bioreactors for wastewater treatment: a review of mechanical cleaning by scouring agents to control membrane fouling. Chem Eng J. https://doi.org/10.1016/j.cej.2016.08.144
10. Behboudi A, Jafarzadeh Y, Yegani R (2017) Polyvinyl chloride/polycarbonate blend ultrafiltration membranes for water treatment. J Memb Sci 534:18–24. https://doi.org/10.1016/j.memsci.2017.04.011
11. Bereskie T, Haider H, Rodriguez MJ, Sadiq R (2017) Framework for continuous performance improvement in small drinking water systems. Sci Total Environ 574:1405–1414. https://doi.org/10.1016/j.scitotenv.2016.08.067
12. Bichai F, Ashbolt N (2017) Public health and water quality management in low-exposure stormwater schemes: a critical review of regulatory frameworks and path forward. Sustain Cities Soc 28:453–465. https://doi.org/10.1016/j.scs.2016.09.003
13. Buonomenna MG (2013) Nano-enhanced reverse osmosis membranes. Desalination 314:73–88. https://doi.org/10.1016/j.desal.2013.01.006
14. Chhipi-Shrestha G, Hewage K, Sadiq R (2017) Microbial quality of reclaimed water for urban reuses: probabilistic risk-based investigation and recommendations. Sci Total Environ 576:738–751. https://doi.org/10.1016/j.scitotenv.2016.10.105
15. da Silva Burgal J, Peeva LG, Kumbharkar S, Livingston A (2015) Organic solvent resistant poly(ether-ether-ketone) nanofiltration membranes. J Memb Sci 479:105–116. https://doi.org/10.1016/j.memsci.2014.12.035
16. Daer S, Kharraz J, Giwa A, Hasan SW (2015) Recent applications of nanomaterials in water desalination: a critical review and future opportunities. Desalination 367:37–48. https://doi.org/10.1016/j.desal.2015.03.030
17. Değermenci N, Cengiz İ, Yildiz E, Nuhoglu A (2016) Performance investigation of a jet loop membrane bioreactor for the treatment of an actual olive mill wastewater. J Environ Manage 184:441–447. https://doi.org/10.1016/j.jenvman.2016.10.014
18. Deshmukh K, Ahamed MB, Deshmukh RR, Pasha SKK, Sadasivuni KK, Polu AR, Ponnamma D, AlMaadeed MA-A, Chidambaram K (2017) Newly developed biodegradable polymer nanocomposites of cellulose acetate and Al_2O_3 nanoparticles with enhanced dielectric performance for embedded passive applications. J Mater Sci: Mater Electron 28:973–986. https://doi.org/10.1007/s10854-016-5616-9
19. Di Vincenzo M, Barboiu M, Tiraferri A, Legrand YM (2017) Polyol-functionalized thin-film composite membranes with improved transport properties and boron removal in reverse osmosis. J Memb Sci 540:71–77. https://doi.org/10.1016/j.memsci.2017.06.034
20. Díez B, Roldán N, Martín A, Sotto A, Perdigón-Melón JA, Arsuaga J, Rosal R (2017) Fouling and biofouling resistance of metal-doped mesostructured silica/polyethersulfone ultrafiltration membranes. J Memb Sci 526:252–263. https://doi.org/10.1016/j.memsci.2016.12.051
21. Drangert J-O, Sharatchandra HC (2017) Addressing urban water scarcity: reduce, treat and reuse—the third generation of management to avoid local resources boundaries. Water Policy wp2017152. https://doi.org/10.2166/wp.2017.152
22. Dulebohn J, Ahmadiannamini P, Wang T, Kim SS, Pinnavaia TJ, Tarabara VV (2014) Polymer mesocomposites: ultrafiltration membrane materials with enhanced permeability, selectivity and fouling resistance. J Memb Sci 453:478–488. https://doi.org/10.1016/j.memsci.2013.11.042
23. Farahbaksh J, Delnavaz M, Vatanpour V (2017) Investigation of raw and oxidized multiwalled carbon nanotubes in fabrication of reverse osmosis polyamide membranes for improvement in desalination and antifouling properties. Desalination 410:1–9. https://doi.org/10.1016/j.desal.2017.01.031
24. Fischer K, Grimm M, Meyers J, Dietrich C, Gläser R, Schulze A (2015) Photoactive microfiltration membranes via directed synthesis of TiO_2 nanoparticles on the polymer surface for removal of drugs from water. J Memb Sci 478:49–57. https://doi.org/10.1016/j.memsci.2015.01.009

25. Gherasim C-V, Mikulášek P (2014) Influence of operating variables on the removal of heavy metal ions from aqueous solutions by nanofiltration. Desalination 343:67–74. https://doi.org/10.1016/j.desal.2013.11.012
26. Gholami A, Moghadassi AR, Hosseini SM, Shabani S, Gholami F (2014) Preparation and characterization of polyvinyl chloride based nanocomposite nanofiltration-membrane modified by iron oxide nanoparticles for lead removal from water. J Ind Eng Chem 20:1517–1522. https://doi.org/10.1016/j.jiec.2013.07.041
27. Ginic-Markovic M, Barclay TG, Constantopoulos KT, Markovic E, Clarke SR, Matisons JG (2015) Biofouling resistance of polysulfobetaine coated reverse osmosis membranes. Desalination 369:37–45. https://doi.org/10.1016/j.desal.2015.04.024
28. Giwa A, Akther N, Dufour V, Hasan SW (2016) A critical review on recent polymeric and nano-enhanced membranes for reverse osmosis. RSC Adv 6:8134–8163. https://doi.org/10.1039/C5RA17221G
29. Giwa A, Akther N, Housani A Al, Haris S, Hasan SW (2016) Recent advances in humidification dehumidification (HDH) desalination processes: improved designs and productivity. Renew Sustain Energy Rev 57:929–944. https://doi.org/10.1016/j.rser.2015.12.108
30. Giwa A, Daer S, Ahmed I, Marpu P, Hasan S (2016) Experimental investigation and artificial neural networks ANNs modeling of electrically-enhanced membrane bioreactor for wastewater treatment. J Water Process Eng 11:88–97. https://doi.org/10.1016/j.jwpe.2016.03.011
31. Giwa A, Dufour V, Al Marzooqi F, Al Kaabi M, Hasan SW (2017) Brine management methods: recent innovations and current status. Desalination 407:1–23. https://doi.org/10.1016/j.desal.2016.12.008
32. Giwa A, Hasan S (2015) Theoretical investigation of the influence of operating conditions on the treatment performance of an electrically-induced membrane bioreactor. J Water Process Eng 6:72–82. https://doi.org/10.1016/j.jwpe.2015.03.004
33. Gkotsis P, Banti D, Peleka E, Zouboulis A, Samaras P (2014) Fouling issues in membrane bioreactors (MBRs) for wastewater treatment: major mechanisms, prevention and control strategies. Processes 2:795–866. https://doi.org/10.3390/pr2040795
34. Gong G, Nagasawa H, Kanezashi M, Tsuru T (2015) Reverse osmosis performance of layered-hybrid membranes consisting of an organosilica separation layer on polymer supports. J Memb Sci 494:104–112. https://doi.org/10.1016/j.memsci.2015.07.039
35. Grandclément C, Seyssiecq I, Piram A, Wong-Wah-Chung P, Vanot G, Tiliacos N, Roche N, Doumenq P (2017) From the conventional biological wastewater treatment to hybrid processes, the evaluation of organic micropollutant removal: a review. Water Res. https://doi.org/10.1016/j.watres.2017.01.005
36. Hasan SW, Elektorowicz M, Oleszkiewicz JA (2014) Start-up period investigation of pilot-scale submerged membrane electro-bioreactor (SMEBR) treating raw municipal wastewater. Chemosphere 97:71–77. https://doi.org/10.1016/j.chemosphere.2013.11.009
37. Huang YW, Wang ZM, Yan X, Chen J, Guo YJ, Lang WZ (2017) Versatile polyvinylidene fluoride hybrid ultrafiltration membranes with superior antifouling, antibacterial and self-cleaning properties for water treatment. J Colloid Interface Sci 505:38–48. https://doi.org/10.1016/j.jcis.2017.05.076
38. Jamaly S, Giwa A, Hasan SW (2015) Recent improvements in oily wastewater treatment: progress, challenges, and future opportunities. J Environ Sci 37:15–30. https://doi.org/10.1016/j.jes.2015.04.011
39. Jiang S, Li Y, Ladewig BP (2017) A review of reverse osmosis membrane fouling and control strategies. Total Environ, Sci. https://doi.org/10.1016/j.scitotenv.2017.03.235
40. Kanagaraj P, Nagendran A, Rana D, Matsuura T, Neelakandan S, Karthikkumar T, Muthumeenal A (2015) Influence of N-phthaloyl chitosan on poly (ether imide) ultrafiltration membranes and its application in biomolecules and toxic heavy metal ion separation and their antifouling properties. Appl Surf Sci 329:165–173. https://doi.org/10.1016/j.apsusc.2014.12.082

41. Katsoyiannis IA, Gkotsis P, Castellana M, Cartechini F, Zouboulis AI (2017) Production of demineralized water for use in thermal power stations by advanced treatment of secondary wastewater effluent. J Environ Manage 190:132–139. https://doi.org/10.1016/j.jenvman.2016.12.040
42. Kim HJ, Choi K, Baek Y, Kim D-G, Shim J, Yoon J, Lee J-C (2014) High-performance reverse osmosis CNT/polyamide nanocomposite membrane by controlled interfacial interactions. ACS Appl Mater Interfaces 6:2819–2829. https://doi.org/10.1021/am405398f
43. Krzeminski P, Leverette L, Malamis S, Katsou E (2017) Membrane bioreactors—a review on recent developments in energy reduction, fouling control, novel configurations, LCA and market prospects. J Memb Sci 527:207–227. https://doi.org/10.1016/j.memsci.2016.12.010
44. Kwan SE, Bar-Zeev E, Elimelech M (2015) Biofouling in forward osmosis and reverse osmosis: measurements and mechanisms. J Memb Sci 493:703–708. https://doi.org/10.1016/j.memsci.2015.07.027
45. Le-Clech P, Chen V, Fane TAG (2006) Fouling in membrane bioreactors used in wastewater treatment. J Memb Sci 284:17–53. https://doi.org/10.1016/j.memsci.2006.08.019
46. Le NL, Nunes SP (2016) Materials and membrane technologies for water and energy sustainability. Sustain Mater Technol 7:1–28. https://doi.org/10.1016/j.susmat.2016.02.001
47. Lee J, Chae HR, Won YJ, Lee K, Lee CH, Lee HH, Kim IC, Lee JM (2013) Graphene oxide nanoplatelets composite membrane with hydrophilic and antifouling properties for wastewater treatment. J Memb Sci 448:223–230. https://doi.org/10.1016/j.memsci.2013.08.017
48. Li Y, Su Y, Zhao X, Zhang R, Zhao J, Fan X, Jiang Z (2014) Surface fluorination of polyamide nanofiltration membrane for enhanced antifouling property. J Memb Sci 455:15–23. https://doi.org/10.1016/j.memsci.2013.12.060
49. Liang S, Qi G, Xiao K, Sun J, Giannelis EP, Huang X, Elimelech M (2014) Organic fouling behavior of superhydrophilic polyvinylidene fluoride (PVDF) ultrafiltration membranes functionalized with surface-tailored nanoparticles: implications for organic fouling in membrane bioreactors. J Memb Sci 463:94–101. https://doi.org/10.1016/j.memsci.2014.03.037
50. Liu C, Lee J, Ma J, Elimelech M (2017) Antifouling thin-film composite membranes by controlled architecture of zwitterionic polymer brush layer. Environ Sci Technol acs.est.6b05992. https://doi.org/10.1021/acs.est.6b05992
51. Liu C, Mao H, Zheng J, Zhang S (2017) Tight ultrafiltration membrane: preparation and characterization of thermally resistant carboxylated cardo poly (arylene ether ketone)s (PAEK-COOH) tight ultrafiltration membrane for dye removal. J Memb Sci 530:1–10. https://doi.org/10.1016/j.memsci.2017.02.005
52. Liu M, Chen Q, Lu K, Huang W, Lü Z, Zhou C, Yu S, Gao C (2017) High efficient removal of dyes from aqueous solution through nanofiltration using diethanolamine-modified polyamide thin-film composite membrane. Sep Purif Technol 173:135–143. https://doi.org/10.1016/j.seppur.2016.09.023
53. Luo W, Phan HV, Xie M, Hai FI, Price WE, Elimelech M, Nghiem LD (2017) Osmotic versus conventional membrane bioreactors integrated with reverse osmosis for water reuse: biological stability, membrane fouling, and contaminant removal. Water Res 109:122–134. https://doi.org/10.1016/j.watres.2016.11.036
54. Ma W, Soroush A, Van Anh Luong T, Brennan G, Rahaman MS, Asadishad B, Tufenkji N (2016) Spray- and spin-assisted layer-by-layer assembly of copper nanoparticles on thin-film composite reverse osmosis membrane for biofouling mitigation. Water Res 99:188–199. https://doi.org/10.1016/j.watres.2016.04.042
55. Maghsoud Z, Pakbaz M, Famili MHN, Madaeni SS (2017) New polyvinyl chloride/thermoplastic polyurethane membranes with potential application in nanofiltration. J Memb Sci 541:271–280. https://doi.org/10.1016/j.memsci.2017.07.001
56. Martín A, Arsuaga JM, Roldán N, de Abajo J, Martínez A, Sotto A (2015) Enhanced ultrafiltration PES membranes doped with mesostructured functionalized silica particles. Desalination 357:16–25. https://doi.org/10.1016/j.desal.2014.10.046

57. McVerry BT, Wong MCY, Marsh KL, Temple JAT, Marambio-Jones C, Hoek EMV, Kaner RB (2014) Scalable antifouling reverse osmosis membranes utilizing perfluorophenyl azide photochemistry. Macromol Rapid Commun 35:1528–1533. https://doi.org/10.1002/marc.201400226
58. Mehrparvar A, Rahimpour A (2015) Surface modification of novel polyether sulfone amide (PESA) ultrafiltration membranes by grafting hydrophilic monomers. J Ind Eng Chem 28:359–368. https://doi.org/10.1016/j.jiec.2015.03.016
59. Mercer KL (2017) 2017 State of the water industry: strengthening our connections. J Am Water Works Assoc 109:56–65. https://doi.org/10.5942/jawwa.2017.109.0090
60. Mi Y-F, Zhao F-Y, Guo Y-S, Weng X-D, Ye C-C, An Q-F (2017) Constructing zwitterionic surface of nanofiltration membrane for high flux and antifouling performance. J Memb Sci 541:29–38. https://doi.org/10.1016/j.memsci.2017.06.091
61. Mohammad AW, Teow YH, Ang WL, Chung YT, Oatley-Radcliffe DL, Hilal N (2015) Nanofiltration membranes review: recent advances and future prospects. Desalination. https://doi.org/10.1016/j.desal.2014.10.043
62. Mokhtari S, Rahimpour A, Shamsabadi AA, Habibzadeh S, Soroush M (2017) Enhancing performance and surface antifouling properties of polysulfone ultrafiltration membranes with salicylate-alumoxane nanoparticles. Appl Surf Sci 393:93–102. https://doi.org/10.1016/j.apsusc.2016.10.005
63. Monisha S, Mathavan T, Selvasekarapandian S, Milton Franklin Benial A, Aristatil G, Mani N, Premalatha M, Vinoth Pandi D (2017) Investigation of bio polymer electrolyte based on cellulose acetate-ammonium nitrate for potential use in electrochemical devices. Carbohydr Polym 157:38–47. https://doi.org/10.1016/j.carbpol.2016.09.026
64. Nebipasagil A, Sundell BJ, Lane OR, Mecham SJ, Riffle JS, McGrath JE (2016) Synthesis and photocrosslinking of disulfonated poly(arylene ether sulfone) copolymers for potential reverse osmosis membrane materials. Polym (United Kingdom) 93:14–22. https://doi.org/10.1016/j.polymer.2016.04.009
65. Neoh CH, Noor ZZ, Mutamim NSA, Lim CK (2016) Green technology in wastewater treatment technologies: integration of membrane bioreactor with various wastewater treatment systems. Chem Eng J 283:582–594. https://doi.org/10.1016/j.cej.2015.07.060
66. Ng LY, Mohammad AW, Leo CP, Hilal N (2013) Polymeric membranes incorporated with metal/metal oxide nanoparticles: a comprehensive review. Desalination 308:15–33. https://doi.org/10.1016/j.desal.2010.11.033
67. Oh HJ, McGrath JE, Paul DR (2017) Kinetics of poly(ethylene glycol) extraction into water from plasticized disulfonated poly(arylene ether sulfone) desalination membranes prepared by solvent-free melt processing. J Memb Sci 524:257–265. https://doi.org/10.1016/j.memsci.2016.11.036
68. Oh HJ, Park J, Inceoglu S, Villaluenga I, Thelen JL, Jiang X, McGrath JE, Paul DR (2017) Formation of disulfonated poly(arylene ether sulfone) thin film desalination membranes plasticized with poly(ethylene glycol) by solvent-free melt extrusion. Polymer (Guildf) 109:106–114. https://doi.org/10.1016/j.polymer.2016.12.035
69. Orooji Y, Faghih M, Razmjou A, Hou J, Moazzam P, Emami N, Aghababaie M, Nourisfa F, Chen V, Jin W (2017) Nanostructured mesoporous carbon polyethersulfone composite ultrafiltration membrane with significantly low protein adsorption and bacterial adhesion. Carbon N Y 111:689–704. https://doi.org/10.1016/j.carbon.2016.10.055
70. Pardeshi PM, Mungray AK, Mungray AA (2017) Polyvinyl chloride and layered double hydroxide composite as a novel substrate material for the forward osmosis membrane. Desalination. https://doi.org/10.1016/j.desal.2017.01.041
71. Prince JA, Bhuvana S, Boodhoo KVK, Anbharasi V, Singh G (2014) Synthesis and characterization of PEG-Ag immobilized PES hollow fiber ultrafiltration membranes with long lasting antifouling properties. J Memb Sci 454:538–548. https://doi.org/10.1016/j.memsci.2013.12.050

72. Pulido BA, Waldron C, Zolotukhin MG, Nunes SP (2017) Porous polymeric membranes with thermal and solvent resistance. J Memb Sci 539:187–196. https://doi.org/10.1016/j.memsci.2017.05.070
73. Qadir D, Mukhtar H, Keong LK (2017) Mixed matrix membranes for water purification applications. Sep Purif Rev 46:62–80. https://doi.org/10.1080/15422119.2016.1196460
74. Qiu G, Ting YP (2014) Short-term fouling propensity and flux behavior in an osmotic membrane bioreactor for wastewater treatment. Desalination 332:91–99. https://doi.org/10.1016/j.desal.2013.11.010
75. Rabiee H, Vatanpour V, Farahani MHDA, Zarrabi H (2015) Improvement in flux and antifouling properties of PVC ultrafiltration membranes by incorporation of zinc oxide (ZnO) nanoparticles. Sep Purif Technol 156:299–310. https://doi.org/10.1016/j.seppur.2015.10.015
76. Razali NF, Mohammad AW, Hilal N (2014) Effects of polyaniline nanoparticles in polyethersulfone ultrafiltration membranes: fouling behaviours by different types of foulant. J Ind Eng Chem 20:3134–3140. https://doi.org/10.1016/j.jiec.2013.11.056
77. Sabir A, Shafiq M, Islam A, Sarwar A, Dilshad MR, Shafeeq A, Zahid Butt MT, Jamil T (2015) Fabrication of tethered carbon nanotubes in cellulose acetate/polyethylene glycol-400 composite membranes for reverse osmosis. Carbohydr Polym 132:589–597. https://doi.org/10.1016/j.carbpol.2015.06.035
78. Safarpour M, Vatanpour V, Khataee A, Zarrabi H, Gholami P, Yekavalangi ME (2017) High flux and fouling resistant reverse osmosis membrane modified with plasma treated natural zeolite. Desalination 411:89–100. https://doi.org/10.1016/j.desal.2017.02.012
79. Shaffer DL, Tousley ME, Elimelech M (2017) Influence of polyamide membrane surface chemistry on gypsum scaling behavior. J Memb Sci 525:249–256. https://doi.org/10.1016/j.memsci.2016.11.003
80. Shafi HZ, Matin A, Akhtar S, Gleason KK, Zubair SM, Khan Z (2017) Organic fouling in surface modified reverse osmosis membranes: filtration studies and subsequent morphological and compositional characterization. J. Memb. Sci. 527:152–163. https://doi.org/10.1016/j.memsci.2017.01.017
81. Shao L, Wang ZX, Zhang YL, Jiang ZX, Liu YY (2014) A facile strategy to enhance PVDF ultrafiltration membrane performance via self-polymerized polydopamine followed by hydrolysis of ammonium fluotitanate. J Memb Sci 461:10–21. https://doi.org/10.1016/j.memsci.2014.03.006
82. Sharma N, Purkait MK (2017) Impact of synthesized amino alcohol plasticizer on the morphology and hydrophilicity of polysulfone ultrafiltration membrane. J Memb Sci 522:202–215. https://doi.org/10.1016/j.memsci.2016.08.068
83. Son M, Kim H, Jung J, Jo S, Choi H (2017) Influence of extreme concentrations of hydrophilic pore-former on reinforced polyethersulfone ultrafiltration membranes for reduction of humic acid fouling. Chemosphere 179:194–201. https://doi.org/10.1016/j.chemosphere.2017.03.101
84. Subramanian S, Seeram R (2013) New directions in nanofiltration applications—are nanofibers the right materials as membranes in desalination? Desalination 308:198–208. https://doi.org/10.1016/j.desal.2012.08.014
85. Tang Y-J, Xu Z-L, Xue S-M, Wei Y-M, Yang H (2017) Tailoring the polyester/polyamide backbone stiffness for the fabrication of high performance nanofiltration membrane. J Memb Sci. https://doi.org/10.1016/j.memsci.2017.07.033
86. Tong T, Zhao S, Boo C, Hashmi SM, Elimelech M (2017) Relating silica scaling in reverse osmosis to membrane surface properties. Environ Sci Technol 51:4396–4406. https://doi.org/10.1021/acs.est.6b06411
87. Tsai H-A, Wang T-Y, Huang S-H, Hu C-C, Hung W-S, Lee K-R, Lai J-Y (2017) The preparation of polyamide/polyacrylonitrile thin film composite hollow fiber membranes for dehydration of ethanol mixtures. Sep Purif Technol 187:221–232. https://doi.org/10.1016/j.seppur.2017.06.060

88. Turek M, Mitko K, Piotrowski K, Dydo P, Laskowska E, Jakóbik-Kolon A (2017) Prospects for high water recovery membrane desalination. Desalination 401:180–189. https://doi.org/10.1016/j.desal.2016.07.047
89. Ulbricht M (2006) Advanced functional polymer membranes. Polymer (Guildf) 47:2217–2262. https://doi.org/10.1016/j.polymer.2006.01.084
90. Vatanpour V, Safarpour M, Khataee A, Zarrabi H, Yekavalangi ME, Kavian M (2017) A thin film nanocomposite reverse osmosis membrane containing amine-functionalized carbon nanotubes. Sep Purif Technol 184:135–143. https://doi.org/10.1016/j.seppur.2017.04.038
91. Velu S, Arthanareeswaran G, Lade H (2017) Removal of organic and inorganic substances from industry wastewaters using modified aluminosilicate-based polyethersulfone ultrafiltration membranes. Environ Prog Sustain Energy https://doi.org/10.1002/ep.12614
92. Verbeke R, Gómez V, Vankelecom IFJ (2016) Chlorine-resistance of reverse osmosis (RO) polyamide membranes. Polym. Sci, Prog. https://doi.org/10.1016/j.progpolymsci.2017.05.003
93. Wang C, Li Z, Chen J, Li Z, Yin Y, Cao L, Zhong Y, Wu H (2017) Covalent organic framework modified polyamide nanofiltration membrane with enhanced performance for desalination. J Memb Sci 523:273–281. https://doi.org/10.1016/j.memsci.2016.09.055
94. Wang H, Wang Z-M, Yan X, Chen J, Lang W-Z, Guo Y-J (2017) Novel organic-inorganic hybrid polyvinylidene fluoride ultrafiltration membranes with antifouling and antibacterial properties by embedding N-halamine functionalized silica nanospheres. J Ind Eng Chem 52:295–304. https://doi.org/10.1016/j.jiec.2017.03.059
95. Wang H, Wei M, Zhong Z, Wang Y (2017) Atomic-layer-deposition-enabled thin-film composite membranes of polyimide supported on nanoporous anodized alumina. J Memb Sci 535:56–62. https://doi.org/10.1016/j.memsci.2017.04.026
96. Wang JJ, Wang Z, Wang JJ, Wang S (2015) Improving the water flux and bio-fouling resistance of reverse osmosis (RO) membrane through surface modification by zwitterionic polymer. J Memb Sci 493:188–199. https://doi.org/10.1016/j.memsci.2015.06.036
97. Wang P, Luo L, Chung TS (2014) Tri-bore ultra-filtration hollow fiber membranes with a novel triangle-shape outer geometry. J Memb Sci 452:212–218. https://doi.org/10.1016/j.memsci.2013.10.033
98. Wang X, Chang VWC, Tang CY (2016) Osmotic membrane bioreactor (OMBR) technology for wastewater treatment and reclamation: advances, challenges, and prospects for the future. J Memb Sci. https://doi.org/10.1016/j.memsci.2016.01.010
99. Wang X, Wang C, Tang CY, Hu T, Li X, Ren Y (2017) Development of a novel anaerobic membrane bioreactor simultaneously integrating microfiltration and forward osmosis membranes for low-strength wastewater treatment. J Memb Sci 527:1–7. https://doi.org/10.1016/j.memsci.2016.12.062
100. Werber JR, Deshmukh A, Elimelech M (2016) The critical need for increased selectivity, not increased water permeability, for desalination membranes. Environ Sci Technol Lett 3:112–120. https://doi.org/10.1021/acs.estlett.6b00050
101. Wu H, Liu Y, Mao L, Jiang C, Ang J, Lu X (2017) Doping polysulfone ultrafiltration membrane with TiO_2-PDA nanohybrid for simultaneous self-cleaning and self-protection. J Memb Sci 532:20–29. https://doi.org/10.1016/j.memsci.2017.03.010
102. Yang B, Yang X, Liu B, Chen Z, Chen C, Liang S, Chu L-Y, Crittenden J (2017) PVDF blended PVDF-g-PMAA pH-responsive membrane: effect of additives and solvents on membrane properties and performance. J Memb Sci. https://doi.org/10.1016/j.memsci.2017.07.045
103. Yang C, Li X-M, Gilron J, Kong D, Yin Y, Oren Y, Linder C, He T (2014) CF4 plasma-modified superhydrophobic PVDF membranes for direct contact membrane distillation. J Memb Sci 456:155–161. https://doi.org/10.1016/j.memsci.2014.01.013
104. Yin J, Deng B (2015) Polymer-matrix nanocomposite membranes for water treatment. J Memb Sci. https://doi.org/10.1016/j.memsci.2014.11.019
105. You X, Ma T, Su Y, Wu H, Wu M, Cai H, Sun G, Jiang Z (2017) Enhancing the permeation flux and antifouling performance of polyamide nanofiltration membrane by incorporation of

PEG-POSS nanoparticles. J Memb Sci 540:454–463. https://doi.org/10.1016/j.memsci.2017.06.084

106. Yu H, Zhang Y, Sun X, Liu J, Zhang H (2014) Improving the antifouling property of polyethersulfone ultrafiltration membrane by incorporation of dextran grafted halloysite nanotubes. Chem Eng J 237:322–328. https://doi.org/10.1016/j.cej.2013.09.094
107. Zhang G, Lu S, Zhang L, Meng Q, Shen C, Zhang J (2013) Novel polysulfone hybrid ultrafiltration membrane prepared with TiO_2-g-HEMA and its antifouling characteristics. J Memb Sci 436:163–173. https://doi.org/10.1016/j.memsci.2013.02.009
108. Zhang H, Li Bin, Pan J, Qi Y, Shen J, Gao C, Van der Bruggen B (2017) Carboxyl-functionalized graphene oxide polyamide nanofiltration membrane for desalination of dye solutions containing monovalent salt. J Memb Sci 539:128–137. https://doi.org/10.1016/j.memsci.2017.05.075
109. Zhang Y, Guo M, Pan G, Yan H, Xu J, Shi Y, Shi H, Liu Y (2015) Preparation and properties of novel pH-stable TFC membrane based on organic-inorganic hybrid composite materials for nanofiltration. J Memb Sci 476:500–507. https://doi.org/10.1016/j.memsci.2014.12.011
110. Zhang Y, Zhao C, Yan H, Pan G, Guo M, Na H, Liu Y (2014) Highly chlorine-resistant multilayer reverse osmosis membranes based on sulfonated poly(arylene ether sulfone) and poly(vinyl alcohol). Desalination 336:58–63. https://doi.org/10.1016/j.desal.2013.12.034
111. Zhao C, Lv J, Xu X, Zhang G, Yang Y, Yang F (2017) Highly antifouling and antibacterial performance of poly (vinylidene fluoride) ultrafiltration membranes blending with copper oxide and graphene oxide nanofillers for effective wastewater treatment. J Colloid Interface Sci 505:341–351. https://doi.org/10.1016/j.jcis.2017.05.074
112. Zhou D, Zhu L, Fu Y, Zhu M, Xue L (2015) Development of lower cost seawater desalination processes using nanofiltration technologies—a review. Desalination 376:109–116. https://doi.org/10.1016/j.desal.2015.08.020

Printed in the United States
By Bookmasters